Public Administration and Information Technology

Volume 26

Series Editor
Christopher G. Reddick, San Antonio, TX, USA

More information about this series at http://www.springer.com/series/10796

Holly Jarman • Luis F. Luna-Reyes
Editors

Private Data and Public Value

Governance, Green Consumption, and Sustainable Supply Chains

Springer

Editors
Holly Jarman
Department of Health
Management and Policy
School of Public Health
University of Michigan
Ann Arbor, MI, USA

Luis F. Luna-Reyes
Department of Informatics
University at Albany
Albany, NY, USA

Public Administration and Information Technology
ISBN 978-3-319-27821-6 ISBN 978-3-319-27823-0 (eBook)
DOI 10.1007/978-3-319-27823-0

Library of Congress Control Number: 2016934042

Springer Cham Heidelberg New York Dordrecht London

Printed on acid-free paper

Springer International Publishing AG Switzerland is part of Springer Science+Business Media (www.springer.com)

Acknowledgments

This book is the result of the collective effort of a research team that collaborated over a period of more than 5 years. The book reflects our current understanding of the importance of technology and governance needed to create applications to open public and private data creating public value. This volume would not have been possible without the hard work and collegiality of the team; the series editor, Christopher Reddick; and the staff at Springer. This project has also received valuable support from a number of students, Weijia Ran, Nic dePaula, Haixin Liu, and James Michaelis. We are grateful to them for their assistance.

We want to especially thank Jana Hrdinova and Mahdi Mirdamadi Najafabadi, whose dedication and tireless effort have been instrumental for the completion of this book. Jana's editorial support along the development of all the book was critical for its completion, and Mahdi's effort during the last stages of the project was also very important for finalizing this volume.

We are also grateful to all our institutions, the Graduate School of Management and George Perkins Marsh Institute at Clark University, the Universidad de las Americas-Puebla, the Center for Technology in Government, the Rockefeller College, and the College of Engineering and Applied Sciences at University at Albany. The in-kind and financial support from these institutions made it possible for us to build on the work of the I-Choose team and to continue our productive collaboration. The research reported in this book was also supported in part by the US-NSF (Grant No. 0955935), CONACYT-Mexico (Grants No. 84082, 117118, and 133670), and the Fulbright Commissions in Mexico and Canada. The views and conclusions expressed in book paper are those of the authors alone and do not necessarily reflect the views of Fulbright, CONACYT, or NSF.

Editors
Holly Jarman
University of Michigan

Luis F. Luna-Reyes
University at Albany

Contents

Authors' Biographies

Mahdi M.N. Abadi Mahdi Mirdamadi NajafAbadi is an informatics PhD student and a graduate of the MPA program at the University at Albany, State University at New York. He has earned his bachelors in software engineering from the University of Tehran (2004), followed by MBA from Sharif University (2007), Tehran. He has over 8 years of experience working in an information security firm and as a management consultant for information technology, organizational development, and business process reengineering. His research interests include government information strategy and management, sociotechnical systems, data analytics, computer modeling, and simulation for education and decision making.

David F. Andersen David F. Andersen is O'Leary and distinguished service professor of public administration, public policy, and information science at the Rockefeller College, University at Albany. His work centers on applying system dynamics, systems thinking, and information technology approaches to problems in the public, not-for-profit, and private sectors, especially using group modeling approaches. He has served as a technical consultant to public and not-for-profit agencies in the federal, state, and local sectors as well as corporate clients in North America and Europe. Most recently he has been working on information standards for bringing sustainable, fair labor, and environmental friendly products to market in the NAFTA region as part of his Carlos Rico Fulbright Award for 2010–2011.

Deborah L. Andersen Deborah Lines Andersen is professor emeritus of information studies and informatics at the University at Albany, State University of New York. She received her PhD from the University at Albany in 1996 and has taught statistics, research design, and information science courses at UAlbany at both master's and PhD levels since then. She served as the codirector of the Information Science PhD program for a number of years, advising cohorts of PhD students in informatics and chairing PhD dissertations in the program. Her research focuses on information policy locally, nationally, and internationally and on public libraries and librarianship. In 2010–2011 she received a Fulbright scholarship to study in Canada and Mexico, concentrating her research on information flows along supply chains and the I-Choose project.

Nicolau DePaula Nic DePaula is a PhD student in information science at the University at Albany, State University of New York. His research addresses issues in the area of e-government, information security, and risk analysis. Mr. DePaula also holds a degree in geography and urban planning from the University at Albany and has worked in multiple IT and research positions for the public and private sector.

Francois Duhamel Francois Duhamel is professor in the Department of International Business in the School of Business and Economics at the Universidad de las Américas Puebla in México since 2006. He received his master's degree in research in strategic management in Paris X-Nanterre and his PhD in business administration with a specialization in strategic management in 2006 from HEC Paris. His main research interests include commercialization of innovations, strategic outsourcing, and international supply chain management. He is a member of the Mexican National Research System. His research has been published in journals such as *European Management Journal*, *Management Decision*, *Technology Analysis & Strategic Management*, *Competitiveness Review*, and *Transforming Government: People, Process and Policy*. He also authored books, books chapters, and conference proceedings articles in those areas.

Isis Gutiérrez-Martínez Isis Gutiérrez-Martínez is a professor in the Department of Business Administration at Universidad de las Américas Puebla in México since 2006. She holds a BS in computer science from Campeche Tech and a MS in business planning and regional development from Mérida Tech in México. She also holds a MS in management of technological and organizational advancements from University of Versailles-Saint Quentin, France, and a PhD in management science from the Conservatoire National des Arts et Métiers in Paris, France. She is a member of the Mexican National Research System. Her research interests focus on human resource management and information technology management. She is an active participant and contributor at international journals and conferences in her field. Her research has been published in journals such as *Government Information Quarterly*, *Competitiveness Review*, and *Transforming Government: People, Process and Policy*. She also has authored books, books chapters, and conference proceedings articles in the aforementioned topics. Her research has received several awards and has been funded by several grants including CONACYT-Mexico (Mexican National Council of Science and Technology) and AECID-Spain (Spanish Agency of International Cooperation for Development).

Jana Hrdinová Jana Hrdinová is an associate director at the Ohio State University Center for Real Estate. Prior to coming to the center, Jana was a program associate at the Center for Technology in Government at the University at Albany, SUNY, where she was responsible for project development, planning, implementation, management, analysis, and writing. Prior to entering the academic world, Jana worked for US Senator (Minnesota) Mark Dayton as deputy director for constituent services concentrating predominantly on immigration issues. In addition to assisting constituents with problems regarding federal agencies, Jana conducted outreach to Minnesota's African-immigrant and African-American communities and advised Senator Dayton on immigration, worker's compensation, and postal service policy issues. Jana holds a master of arts in political science and a bachelor of arts in international relations from the University of Minnesota.

Holly Jarman Holly Jarman is a political scientist who studies the effects of market regulation, particularly cross-border regulation and international trade and investment agreements, on health and social policies. Within this broad topic, her publications address questions related to the relationship between tobacco control regulation and the world trading system, the regulation of cross-border health markets in the European Union, cross-border collaborations to promote innovative public health and environmental regulation, and the impact of cross-border patient mobility on global health governance. She received her PhD in political science from the London School of Economics and Political Science (LSE) for comparative research into the use of non-trade issues, including environmental protections and labor standards, as weapons and incentives in EU and US trade policy.

Joanne Luciano Dr. Joanne Luciano is the president and founder of Predictive Medicine, Inc., a Boston-based research and development consultancy, and a visiting research scholar at the Center for Cognition, Communication, and Culture at Rensselaer Polytechnic Institute. Dr. Luciano is a pioneer in the use of computational methods in clinical research in psychiatry, the application of semantic technologies to healthcare and the life sciences, and interoperable data architecture for certification and inspection to support sustainable consumption and supply chains that build trust among public and private actors. An internationally recognized multidisciplinary scientist, her research focuses on social and computational technologies that enable the integration and reusability of research artifacts. She cofounded the field of Health Web Science, which provides a foundation to study the role of the Web in all aspects of the healthcare ecosystem. Luciano is currently working to promote the adoption of the Financial Industry Business Ontology (FIBO), which provides a common conceptual model to enhance interoperability, transparency, risk exposure assessment, and regulatory compliance across the financial industry worldwide.

Luis F. Luna-Reyes Luis Felipe Luna-Reyes is an associate professor of informatics at the University at Albany in Albany, NY. He holds a PhD in information science from the University at Albany, and he is also a member of the Mexican National Research System. His research focuses on electronic government and on modeling collaboration processes in the development of information technologies across functional and organizational boundaries. His research interests are related to areas such as interorganizational collaboration, information sharing, success of government-wide Web sites, and information policy to promote economic exchange in the

NAFTA region. He is the author or coauthor of articles published in *Government Information Quarterly*, *European Journal of Information Systems*, *International Journal of Electronic Government Research*, *Gestión y Política Pública*, and *System Dynamics Review*, among others.

Theresa A. Pardo Theresa A. Pardo is the director of the Center for Technology in Government, University at Albany, State University of New York (SUNY), and a research associate professor at the Rockefeller College of Public Affairs and Policy and the College of Computing and Information. Dr. Pardo serves as Open NY Policy adviser to New York State's Governor Andrew Cuomo, as a member of the National Advisory Committee for the US Environmental Protection Agency, and as president of the Digital Government Society. Her work has been funded by the US National Science Foundation (NSF), the Social Science and Humanities Research Council (SSHRC) of Canada, the Library of Congress, the World Bank, the United Nations, Microsoft, SAP, and numerous governments at all levels, among others.

Sergio Picazo-Vela Sergio Picazo-Vela is a professor in the Business Administration Department at the Business School of the Universidad de las Américas Puebla, and he is member of the Mexican Research System. He received his PhD in business administration from Southern Illinois University, Carbondale. His research interests include e-government, e-commerce, information technology outsourcing, and social media sites. His research has been published in different peer-reviewed journals such as *Government Information Quarterly*, *Communications of the Association for Information Systems*, *Computers in Human Behavior*, and *Competitiveness Review*. His research has also been published in proceedings of different conferences such as the Annual International Conference on Digital Government Research, Annual Meeting of the Decision Sciences Institute, and AMCIS Doctoral Consortium. Dr. Picazo-Vela has participated writing chapters for books published by Mexican and International Publishers and writing opinion articles for digital and traditional newspapers. Finally, he has experience developing IT solutions to create value at organizations.

Weijia Ran Weijia Ran is a PhD candidate in informatics at the University at Albany. Her research interests include sociotechnical systems, data management, analytic-based decision making, modeling and simulation, research methodologies, and sustainability.

Djoko S. Sayogo Dr. Sayogo is senior lecturer in the Department of Economics and Business at the University of Muhammadiyah at Malang (UMM), Indonesia. Djoko currently served as the director at the Center for Economics, Business and Entrepreneurships Development at the Department of Economics, University of Muhammadiyah at Malang. Djoko works on the CTG team that aims to develop a proof of concept for global information systems using semantic Web technologies based on OWL ontology, called I-Choose. He is a former Fulbright presidential scholar from Indonesia.

Giri K. Tayi Giri Kumar Tayi is a professor of management science and information systems at the State University of New York at Albany. His research and teaching interests are interdisciplinary and span the fields of information systems, operations management, and operations research. He serves or has served as SE and AE on the editorial boards of academic journals such as *Information Systems Research*, *IEEE Intelligent Systems*, *IEEE Computing Now*, *Decision Sciences*, *ACM Journal of Data and Information Quality*, *Information Technology Management*, *Information Systems Frontiers*, *Information and Management*, and *International Journal of Shipping and Transport Logistics*. He has over 55 refereed publications and 65 conference proceedings. His papers have appeared in *Operations Research*, *ISR*, *Management Science*, *MIS Quarterly*, *Quantitative Marketing and Economics*, *IEEE Transactions*, *Networks*, *Naval Research Logistics*, *EJOR*, *Journal of Combinatorial Optimization*, *INFORMS Journal on Computing*, *Journal of Computer Security*, *Government Information Quarterly*, and *Communications of the ACM*.

Jing Zhang Jing Zhang is an associate professor of management at Clark University. Her research interests focus on interorganizational information and knowledge sharing, organizational impact of technology and innovation, and the role played by information technology to support sustainable consumption and sustainable supply chain. Her work is supported by National Science Foundation and Mosokowski Institute of Public Enterprises. Her publications appear in *European Journal of Information Systems*, *Government Information Quarterly*, *Public Performance and Management Review*, and *Information Technology and Management*, among others. Professor Zhang holds a PhD in information science from the University at Albany, State University of New York.

Chapter 1
Public Value and Private Organizations

Holly Jarman, Luis F. Luna-Reyes, and Jing Zhang

Abstract The first chapter of the book introduces our key question: how can private actors be incentivized to share their data in a way that promotes the public value of the information disclosed? We are interested in whether and how these different organizations can be encouraged by governments and other interested actors to share the information that they hold. The means by which this might be accomplished—particularly how these private actors might be encouraged to collaborate among themselves and with governments—is a major focus of the book. The chapter explores the concept of public value in the context of data disclosure by private organizations, using empirical evidence from the I-Choose project. We argue that while disclosing product information can enhance the public sphere, information disclosure alone is not enough to guarantee this. Disclosure must be supported by innovative governance mechanisms. The chapter explains why disclosing private product data is considered valuable by some policymakers and advocates and considers the barriers to disclosing product information.

Keywords I-Choose • Public value • Smart disclosure • Open data

1.1 Introduction: The Puzzle

As our ability to electronically collect, manipulate, and publish large amounts of data increases, making private information available to the public is increasingly a viable part of what governments can and should do to fulfill their mandates.

H. Jarman (✉)
Department of Health Management and Policy, School of Public Health,
University of Michigan, Ann Arbor, MI, USA
e-mail: hjarman@umich.edu

L.F. Luna-Reyes
Department of Informatics, University at Albany, Albany, NY, USA
e-mail: lluna-reyes@albany.edu

J. Zhang
Graduate School of Management, Clark University, Worcester, MA, USA
e-mail: jizhang@clarku.edu

H. Jarman, L.F. Luna-Reyes (eds.), *Private Data and Public Value*,
Public Administration and Information Technology 26,
DOI 10.1007/978-3-319-27823-0_1

There is much current interest in the power of information disclosure to improve our lives: calls for "open government" argue that the routine disclosure of information about how government works and what it is doing can support more effective public oversight. Discussions about the power of "big data"—very large datasets that require innovative methods of processing, curation, and dissemination—often focus on ways in which big data might be used to tackle intractable problems in our society such as crime or pollution. "Smart disclosure" policies, promoted by the Obama Administration in the United States, promise to deliver benefits for the society by disclosing information held on individual citizens or consumers back to the source.

These debates have implications for many areas of policy, including healthcare and public health, energy and the environment, and banking and finance. Governments have taken some important first steps by adopting open government strategies and policies; by fostering research on the management, dissemination, and interoperability of big data; and by applying the principles of "smart disclosure" to their own records.

But a great deal of valuable information that could be used to solve societal problems is not held by governments—it is held by private organizations. It is not publicly available and is often held in proprietary systems. It may, in fact, be a commercial secret. Collecting this information bears a cost, and revealing it to others may be perceived as costly as well. In the case of environmental sustainability, for example, governments hold information about emissions and pollution. They can disclose information such as which firms have broken the rules on emissions or pollution, which sites are polluted, and what funds have been dispersed as part of government-funded cleanup initiatives. But when it comes to individual consumer choices that affect the environment, businesses hold much of the information that consumers need in order to make decisions—particularly about factors that consumers cannot directly observe, such as how "green" a firm's production process is.

Many argue, therefore, that there are important benefits to disclosing privately held data. First, disclosure—if done correctly—should allow individual members of the public better access to, and control over, their own data. For instance, patients would be able to access their own health data, or the customers of energy companies would be able to review their energy consumption.

Second, it is argued that individuals will act on this new information, making their actions more efficient or valuable for either themselves, the public, or both (Sunstein, 2012). Upon viewing their health data, patients might have a better sense of their own health and choose to make positive changes in their lifestyle. Energy company customers might find ways to reduce their electricity consumption. Having access to production practices, consumers may choose to buy products that respond to their values related to health and economic or environmental sustainability.

Third, it is argued that better aggregate public awareness and scrutiny of what governments, businesses, and other organizations do, including public interest organizations and the media, will lead to demands for better behavior. On viewing their health data more regularly, patients might become aware of a need to reduce healthcare costs or to obtain more comprehensive health insurance coverage.

On monitoring their energy usage, customers might demand more information about renewable sources of electricity or regulation that makes it easier for them to switch providers. Public awareness of production practices in the food or fashion industries may promote more responsible corporate practices.

Finally, policymakers often argue that disclosure will lead to increased opportunities for innovation and economic growth based on the disclosed data—that new industries will emerge to filter and process the data on behalf of consumers. This can be seen in the many small online services created in 2013 promising to help individuals and families in the United States navigate new health insurance portals that present the users with large quantities of information about insurance plans. It can also be seen in new information intermediaries like GoodGuide, which offers consumers information about product impacts on health, the environment, or society.

This leads to our key question: how can we incentivize private actors to share their data in a way that promotes the public value of the information disclosed? This question needs unpacking a little. Our category of "private actors" involves not only businesses and industry but also nongovernmental organizations (NGOs). We are interested in whether and how these different organizations can be encouraged by governments to share the information that they hold. The means by which this might be accomplished—particularly how these private actors might be encouraged to collaborate among themselves and with governments—is a major focus of this book. Finally, although we choose to remain optimistic about the potential for information disclosure as a policy tool, we also point out weaknesses in the assumptions that lie behind such disclosure policies. Disclosure does not guarantee that public or firm behavior will change, and the effectiveness of disclosure policies is sensitive to the governance tools that are employed in parallel.

We argue in this book that disclosing privately held data can have a public value under certain conditions. In our definition, a policy outcome that has public value satisfies two criteria. It is (1) in line with what adds value to the public sphere, (2) as determined by a fair and transparent consultative process that aligns agency goals with democratic mandates. (This definition is unpacked further in Sect. 1.3.) Our book explores the necessary and desirable conditions under which information disclosure can produce public value—in particular, we discuss the governance conditions and mechanisms required to produce policy outcomes with public value. We then explore the potential for connecting this public value with the interests of private organizations, explain the roadblocks to doing so, and outline some first steps for overcoming them.

We approach our core question through a case study of coffee supply chains in North America. Coffee is a commonly traded product that seems relatively simple, yet has a number of complex supply chains. We focus on coffee that is grown in Mexico and sold in Canada and the United States, three countries that are governed by a shared trade regime but with very different regulatory environment, governance styles, levels of economic development, and patterns of information technology use. Importantly for our purposes, coffee is often connected to key, nonpecuniary values, such as environmental sustainability and worker and human rights. Coffee is available in many different types—the bag of coffee you buy in the store might claim to be organically or shade-grown or might be certified as "fair trade." In practice, there is no feasible way for the consumer to verify these claims—they must rely on truthful reporting by producers, distributors, retailers, and certifiers.

We use our coffee case study to think about how private data disclosure might help make supply chains more sustainable. We explore to what extent disclosing information about how coffee was produced using emerging technologies can support green purchasing decisions by consumers and ultimately change the behavior of other actors in the supply chain. We find that governments cannot take a backseat in this process. In order for consumers to trust the data that is disclosed, we need to formulate new governance mechanisms that are a good fit with the age of big data. Disclosure alone will not suffice—creating more sustainable supply chains requires addressing classic questions of governance about the independence of organizations, accountability, and transparency.

The next sections in this chapter discuss these core concepts and debates in more detail. First, we introduce our case study—the I-Choose project—as an example of private data disclosure with the potential to create public benefits. Second, we explore the concept of public value in the context of a vibrant scholarly debate about its definition and uses. In particular, we focus on what public value means in the context of data disclosure and relate it to the use of new and emerging technologies. Third, we discuss reasons why private organizations in the supply chain might choose to disclose data, before discussing barriers to such disclosure. Finally, we summarize our argument and lay out the road map for the rest of the book.

1.2 The I-Choose Project

The findings and conclusions presented in this book are drawn from a National Science Foundation funded study that examined the coffee supply chain to better understand the requirements and impacts of information disclosure on firm and consumer behavior. Our interdisciplinary and international research team combined lessons from the disciplines of information science, management, political science, business, and computer science to address this question. Although this study focused on one consumer product, we believe that the lessons we draw from it are broadly applicable, with implications in areas such as environmental policy, healthcare, trade, finance, and food safety.

Our project focused on one commodity—coffee that is grown in Mexico and sold in Canada and the United States. Coffee itself seems like a simple product—because it is a commodity rather than a product with parts or ingredients sourced from many different countries. A bag of coffee beans bought in your local store has a relatively simple set of possible supply chains. Growers cultivate coffee plants, intermediary organizations process coffee cherries into green coffee beans (in Mexico, growers are often organized into coffee cooperatives), and exporters sell green beans to roasters (sometimes through brokers). Roasters—from large multinational firms such as Nestle to small independent businesses—turn green beans into the coffee beans that we recognize, roasting and packaging them and selling them to retailers, from large supermarket chains to smaller independent stores. However, how coffee is cultivated, produced, distributed and sold raises important social and environmental questions, answers to which are often not visible to consumers buying coffee in stores.

Coffee is one of the main crops in Mexico. Given its production volume as well as the income derived from its export, coffee is a strategic crop for the country (SAGARPA, 2012). Mexico has over 280,000 coffee producers, of which over 200,000 are smallholders—small farms producing a mixture of crops for consumption and income (Fridell, 2007). The United States is the main market for green coffee grown in Mexico, with 70 % of coffee beans being grown for export. Ninety-eight percent of the coffee produced in Mexico is the higher-quality Arabica coffee, with the rest being robusta coffee (USDA Foreign Agriculture Service, 2011).

Production for export can be damaging, and variations in the world price of coffee can have serious social consequences. Some of the poorest areas of Mexico with some of the largest indigenous populations, such as Chiapas and Oaxaca, are Mexico's most significant coffee producers. Due to more industrialized farming methods and planting programs sponsored by major international companies, global coffee production is now often greater than global consumption, causing prices to drop. Not only that, but when global demand for cheaper robusta coffee increases, the price of the more expensive, better-quality Arabica beans grown by most Mexican producers drops. When the price of Arabica drops, Mexican coffee growers suffer. Coffee prices can be very volatile, but because of the time, investment, and labor involved in creating a coffee plantation, growers cannot easily switch to another source of income, at least not without losing their entire investment.

The result of this instability can be devastating for the communities where coffee is produced. Mexico's decision to sign the North American Free Trade Agreement in 1995 has had the effect of increasing the ability of large multinational businesses to establish themselves in Mexican markets. In agriculture, large international agribusiness and processed food retailers often displaced local growers in serving domestic consumers. Many coffee growers stopped producing mainly for domestic markets and now produce mainly for export. The migration caused by the reconfigured North American business environment—internal, from rural areas to cities, and external, from Mexico to the United States—has had disrupting effects on the labor supply that coffee growers rely on to harvest their crops.

Coffee cultivation raises environmental concerns, too. The creation of industrial coffee plantations can result in widespread deforestation with subsequent consequences for soil quality and the diversity of flora and fauna, threatening populations of insects and birds. But more traditional forms of growing coffee as part of the natural forest ecosystem, also called shade-grown, do not involve such deforestation and are seen as highly beneficial to biodiversity conservation in tropical forest ecosystems (Rice, Ward, Smithsonian Migratory Bird Center, & Natural Resources Defense Council, 1996). Industrial coffee plantations are also more likely to use large quantities of chemicals—such as pesticides and fertilizers—in growing their coffee. These chemicals can contaminate the soil, further disrupt ecosystems, and find their way into the groundwater with real consequences for local communities.

Because of the social and environmental consequences of coffee production, alternative manufacturing procedures and trading systems have emerged that try to address these problems. Coffee can now be bought in several different varieties that signal that its production followed certain societal and/or environmental values. Coffee can be certified as "fair trade," meaning that in the least it guarantees a minimum price to producers that is above the price that a completely open market would provide, and at most the pledge that the coffee was produced in a way that was sustainable and ethical, and invests in the communities that created it. Coffee can also be labeled as "organic," indicating that it was produced without using harmful chemicals, or "shade-grown," meaning that it was produced via traditional methods that avoid deforestation rather than industrial farming methods. The stated purpose of many of these certification systems is to create a more sustainable coffee trade that protects the environment, provides sustainable incomes, and allows coffee-growing communities to flourish.

A significant proportion of Mexican coffee cooperatives produce coffee that falls under one or more of these categories. Mexico is the world's largest producer of organic coffee, using 10 % of the land to produce this category of coffee (SAGARPA, 2012). Eighty-five percent of organic coffee produced in Mexico is intended for export, with most of the organic coffee in Mexico produced in Oaxaca and Chiapas (USDA Foreign Agriculture Service, 2011). On the other hand, Mexico was instrumental in the creation of the first fair-trade seal, Max Havelaar, in 1988, and has played a key role in promoting these practices since then through large cooperatives such as UCIRI (Fridell, 2007). Additionally, there are currently about 37,500 acres of land producing Rainforest Alliance certified coffee (Rainforest Alliance, 2015). This amount is expected to increase in an important way by 2020 because of current partnerships between the Rainforest Alliance and Nestlé as part of the Nescafé plan (Nescafe n.d.; Rainforest Alliance n.d.).

These differences in how coffee can be produced, and how consumers perceive them, lie at the core of the I-Choose project. Customers can now go into most big grocery stores and buy coffee labeled as fair trade. But the "fairness" of that coffee, the criteria required for it to receive its fair-trade label, can vary a great deal. The customer may want to know information about how the coffee was produced and distributed, but cannot directly observe these product characteristics (see Chap. 2 in this volume and also Sayogo, Zhang, Pardo et al., 2014; Sayogo et al., 2015; Sayogo et al., Forthcoming; Sayogo, Zhang, Liu, Picazo-Vela, & Luna-Reyes, 2014). The consumer most likely ends up making a choice based on what they can observe—the price—and what they see on the packaging. However, a label stating that a coffee product is "ethical" or "green" in some way may just be an assertion made by the seller and need not come with any explanation as to what the label really means.

Coffee production and distribution in North America therefore gives us a relatively simple product to study, but one with enormous implications for the sustainability of the environment and communities. We use this case to explore our key question: how can we incentivize private actors to share their data in a way that promotes the public value of the information disclosed? In the context of our coffee case study, this main question breaks down as follows: we explore the idea that the disclosure of information about how coffee is produced, distributed, and sold can influence the purchasing decisions of consumers and the behavior of producers, distributors, and retailers themselves. In this volume, we are most interested in the governance mechanisms that are necessary and desirable not only to promote disclosure but to produce trusted data that creates public value.

The evidence used in this book comes from multiple sources, including interviews and focus group feedback from key stakeholders, government documents, scholarly literature, and coffee certification and inspection data. More specifically, the project started with workshops involving a group of stakeholders in the coffee supply chain. We continued our exploration with a systematic review of some of the most important certification schemes for organic and fair-trade products, as well as with a series of interviews with many other supply chain participants in Mexico and the United States. Finally, our research included building a data architecture proof of concept and ontology development using semantic web technologies (see Methodological appendix).

Using the evidence gathered from the I-Choose project, we argue that disclosing privately held data can create public value—providing that the disclosure is supported by appropriate governance mechanisms. The next section unpacks the concept of public value and discusses what it means in the context of new and emerging web technologies.

1.3 Public Value

Public value is a term with many definitions. We take a broad approach to public value that is not just about direct public management of service provision but which emphasizes how governments might act to incentivize private actors. This goes against narrower definitions of public value that envision governments and individual consumers in a producer-customer relationship. Instead, we note the value of bringing information into the public sphere, where it can be scrutinized, assessed, and used to uphold the public interest. We emphasize a range of public, consensual "values" that are not just about delivering economic efficiency or the cheapest product (Jorgensen & Bozeman, 2007). And we recognize the potential for private organizations to uphold public values through their own actions.

The concept of "public value" was first put forward by Mark Moore in his book *Creating Public Value* (Moore, 1995). Designed as a practical toolbox for public managers, the book outlined an approach to public management that could be used to orient public bodies more firmly toward the needs and desires of citizens and stakeholders rather than just the agency's own needs or those of the government hierarchy.

Moore's book addresses a core problem that is commonly outlined in theories of public administration and public policy: government agencies have a tendency to want to deliver things to their own advantage (Dunleavy, 2002; Wilson, 1886), or to the advantage of key rent seekers (Niskanen, 1994), rather than to the advantage of the public. In the first scenario, government bodies are viewed as competitive actors that seek autonomy, prestige, and resources, competing against other government agencies to get them. The second scenario, taken to its logical conclusion, results in government agencies that are captured by powerful interest groups, designing policies and regulations to serve these groups rather than the interests of the public as a whole. Underpinning this scenario is the idea that those who shout the loudest have the most influence—concentrated interests with access to resources are presumed to have a greater impact upon policy decision-making than the much more diffuse public interest (Olson, 1965; Schattschneider, 1935).

Subsequent scholarship has shown that the reality of preference formation and decision-making in government agencies is far more complex. Bureaucrats and public managers have their own preferences, while at the same time, many consider themselves public servants with a duty to support the public interest (Paige, 1997). Moore's discussion of public value supports this view. His book examines the ways through which public agencies might be incentivized to deliver on goals that represent the collective, public interest, as well as concrete ways to formulate what public value means in different settings and measure progress toward achieving policy outcomes that uphold public value as defined.

After the initial introduction of public value, however, the concept began to be more widely used—and sometimes abused—being narrowly interpreted as a kind of customer satisfaction criterion for governments. Some scholars raised questions about the usefulness of such a slippery concept. For example, perhaps the simplest definition of public value is "what the public values" (Benington & Moore, 2011). But finding out what the public values is not easy and what the public values is not always coherent. For example, public opinion polls demonstrate that we as citizens are capable of simultaneously valuing extensive public services and low taxation (Smith, 2015). Influenced by the media, public opinion can change rapidly, responding to headline issues rather than long-term problems. And so, following explicitly what the public wants at any one time does not necessarily make for stable policy or good government.

As the concept of public value became more narrowly defined in terms of customer service, Bennington and Moore (2011) and Moore (2013) attempted to reclaim it by exploring more fully its theoretical underpinnings. In particular, scholars such as Bozeman (2007) and Moore (2013) pushed back against the New Public Management (NPM) paradigm, stating that the core ideas behind the public value perspective, particularly public action in pursuit of collective social values, were in stark opposition to NPM. In NPM, the individual preferences of citizens are held to be very important, with public service improvements driven by the choices of citizen-consumers. Although both public value and NPM approaches claim to be more "customer facing" than previous public administration paradigms, the "customers" in each case are very different. From Moore's version of the public value perspective:

> …the relevant "customer" is a collective public (local, regional, or national) acting through the imperfect processes of representative democracy rather than an individual consumer making choices about what to buy for personal benefit. (Moore, 2013: 3, emphasis added)

By connecting public value more closely to representative democracy, Moore therefore envisions a balance for public managers in delivering upon democratic mandates and consulting with stakeholders. Finding a balance is complex given the diversity of values, which go well beyond the economic ones. Beck Jørgensen and Bozeman (2007), for example, identified seven constellations of public values emerging from the interactions among politicians, public managers, citizens, the environment, and the society at large. These constellations include values such as the protection of minorities, shareholder value, dialogue, governance, or citizen involvement, just to mention few of them (Abolafia, 2001; Bozeman, 2007; Michalos, 2008; Moore, 1995).

Current debates surrounding the definition of public value recognize this difficult balance and build upon this simple definition in order to provide a much more nuanced description of public value. Drawing from this literature, despite its diversity, we can discern some common elements:

- *Public value is a concept tied to anapproach.* The process through which public value is sought is just as important as the definition of public value itself. In Moore's (1995) original terms, values constitute strategic outcomes that require the appropriate operational capabilities and the stakeholder engagement necessary for support and legitimacy of the policy mechanisms. Alignment between public goals and policy outcomes valued by the public is credited by some with increasing trust in government and therefore enhancing government legitimacy (Center for Technology in Government, 2011).
- *Public value approaches are strategic and longterm.* Benington and Moore's (2011) broader definition implies that the concept of public value goes beyond any one administration or set of institutions to provide value not just to the current public but to society as a whole and even future generations. A public value approach is frequently described as "strategic," meaning that it is a longer-term approach that is more than just a series of knee-jerk reactions to public or stakeholder demands.
- *Public value approaches should beparticipatory.* In a public value approach, what is valuable is agreed upon through a participatory and collaborative process as opposed to a one-way process where government is "informed" of stakeholder preferences. The idea is that public value should be "cocreated" through a two-way process (Benington & Moore, 2011, p. 50). This participatory process should satisfy key democratic criteria. It should be seen to be legitimate, transparent, and inclusive. Bennington, for example, argues that public value underpins an emerging paradigm of "networked community governance," a shift toward emphasizing the role of civil society over that of the state and markets. In networked community governance, the dominant form of control is not bureaucratic hierarchy or market forces but networks and the norms that they propagate (Benington, 2011; Stoker, 2006).
- *Public value approaches aim to enhance the publicsphere.* The process of creating public value therefore focuses on policy outcomes that the public collectively values, takes a long-term view, and is participatory. These features, taken

together, account for the final characteristic of public value: public value approaches aim to enhance the public sphere. The public sphere can be defined as a conceptual space for public discussion and debate that allows political discourse, including critique of public authority (Calhoun, 1992; Habermas, 1989). The ideal public value approach, therefore, attempts to add value to the public sphere (Benington, 2011). It does this by creating a robust process for the deliberation of how governments should act upon democratic mandates—a process that encourages participation from nongovernmental organizations, aims to produce policy outcomes that the public collectively values, and takes a long-term approach to delivering on those goals.

This discussion of the public sphere is especially interesting when we consider the role of new and emerging technologies, particularly web technologies. As many scholars have pointed out (Castells, 2007, 2012; Fernback & Thompson, 1995; Rheingold, 2008), the most utopian depictions of the web view it as extending the public sphere—allowing more people access to more information than ever before and creating virtual spaces through which we can debate and deliberate key issues. Proponents of this view argue that the web increases transparency regarding what governments, corporations, or other organizations do and how they do it. It allows previously secret information to be available to mass public, allowing greater public deliberation and participation in decision-making. Research has shown that the concept of public value is central to understanding how open government policies can support a range of desirable social, economic, or other policy goals but that policymakers attempting to maximize the public value of open government policies should collaborate in order to define what public value means in each context and how it might best be achieved (Center for Technology in Government, 2011).

Critics of this view point out that the reality of the web has turned out quite differently—they argue that the scrutiny and participation that web optimists hope for will not occur automatically. They raise concerns that the web has become a commercialized space where powerful actors can tightly control the messages that they send and misinformation abounds. Some argue[1] that the web actually weakens the public sphere, giving members of the public the impression that they are being consulted, while preserving existing inequities in power and resources (Boeder, 2005). Others point out that increased transparency works both ways, tempting governments and other organizations to conduct mass surveillance and collect large amounts of personal data as evidenced in the surveillance of US citizens' internet and phone records conducted by the US National Security Agency following 9/11 (RussiaToday, 2013). See also Fuchs, Boersma, Albrechtslund, and Sandoval (2012) as well as news and comments related to the collection of email communication gathered by NSA in the last years (see http://rt.com/usa/nsa-internet-terrorism-years-810/).

What should we learn from this debate for our definition of public value? The critiques of the web as a public sphere are important because they indicate that

[1] Including Habermas himself.

disclosure alone is not enough to deliver public value by our definition. It is not enough to disclose information about what governments and businesses do. That information has to be distributed, filtered, analyzed, and ultimately shown to be trustworthy if it is to add value to the public sphere. In other words, information disclosure has to be supported by a trusted governance process.

The next sections explore the use of information disclosure as a policy tool, building on our definition of public value to consider the contribution of information disclosure to the public sphere. Focusing more tightly on our case study, we first explain why disclosing private product data is considered valuable by some policymakers, advocates, and firms, before discussing potential barriers to data disclosure.

1.4 Why Disclose Private Product Data?

The disclosure of private product data is argued to produce three distinct categories of benefit for the public:

- *Citizen empowerment*, allowing individuals better access to and control over their own data
- *Public scrutiny*, better public awareness and scrutiny of what governments and other organizations do, leading to demands for better policies
- *Innovation and growth*, increased opportunities for innovation and economic growth based on the disclosed data, with the assumption that this growth will be passed on in ways that benefit the wider public

Citizens empowered by information gained through disclosure might be incentivized not only to take steps to improve their individual situation, such as decreased energy usage, changes in health behavior or diet, or better financial planning, but might also be influenced to choose products and services that are better aligned with their ethical values. Likewise, better public scrutiny might boost corporate social responsibility, not just trust in public authorities. And for private companies, disclosure may enable them to differentiate their products within a crowded marketplace, earning them a price premium.

The disclosure of product data held by public, private, and nongovernmental organizations has the potential to benefit those organizations as well as consumers accessing the disclosed data. When thinking about product data disclosure, or private sector transparency more generally, it is equally important to think about the perceived commercial value of the data to be disclosed. The central goal of any policy or mechanism encouraging private sector transparency should be to facilitate this alignment. This means ensuring that brand value can be maintained or enhanced through product data disclosure initiatives and that any such scheme offers value to business in differentiating their products within crowded markets. The following considers these public and private benefits in turn.

1.4.1 Benefits to Public Agencies

In raising the issue of data disclosure, policymakers have three discrete but interconnected goals: to promote greater consumer access to information which can influence the goods and services that they purchase; to promote innovative use of data in ways that can increase profitable economic activity, such as providing services to aid consumer choice; and to reduce regulatory burdens and costs through greater data transparency and public-private collaboration (Executive Office of the President, 2013).

Access to Information First, policymakers wish to increase consumers' access to information about the products that they buy. Consumers currently have far less access to product information than other actors in the supply chain such as producers and retailers. Outside of the observable characteristics of the product, consumers must rely heavily on producers to provide them with information to assist their decisions, resulting in suboptimal decision-making (Akerlof, 1970). Information asymmetries tend to decline over time as markets grow and mature and as information about product quality is acquired through repeated purchases (Wankhade & Dabade, 2006). Some products, however, such as tomatoes, coffee, or beef, are less likely to experience such dynamics because they are not frequently linked to information other than price.

To counteract this information asymmetry, a growing number of consumers are turning to new technologies to determine information about product characteristics that are not directly observable, such as the distance the item has traveled, the chemicals used in its production, or the labor conditions under which the product was manufactured. We already provide some assistance to consumers through mandatory product labeling which requires manufacturers to list ingredients and calorie counts, for example. But there is a growing pressure among consumers to expand the range of information that they can access regarding their purchases and increased recognition among companies that wish to be "socially responsible" that disclosing such information can be to their advantage.

Innovation and Growth Second, policymakers want to promote the creation of innovative consumer products as a means of increasing economic growth. It is envisioned that making more data public in reusable formats will promote its use in a range of new ways, including applications available to consumers about the products and services that they purchase. If the data disclosed is of high enough quality, it is likely that developers and entrepreneurs will be keen to use it in their future projects. Organizations and individuals may well be willing to pay a premium in order to access the information in a form that is convenient and relevant to their needs.

Reduce Costs Third, policymakers seek to reduce regulatory burdens and costs through greater data transparency and public-private collaborations. In an era when public attention is highly focused on budgetary constraints, governments are looking for innovative ways to save money. One way to do this is to change how the costs of regulation are distributed among taxpayers, businesses, and individuals acting as consumers.

At its core, this strategy relies on increasing the transparency of information not just about products but about the private sector as a whole. Many authors have discussed requirements for opening data (Executive Office of the President, 2013; Kalampokis, Tambouris, & Tarabanis, 2011; Lourenço, 2015; Zuiderwijk, Jeffery, & Janssen, 2012). On the basis of this research, transparent data can be defined as data that is (1) publicly available, (2) easily understood by nonexperts, (3) published in an accessible format, (4) via accessible media, and (5) released on a timely schedule. The concept of private sector transparency, which can be linked back to more inclusive definitions of stakeholder, has been evolving over time, from the confrontational stakeholder tactics to partnerships and collaborative approaches where technology can play a key role (Baue & Murninghan, 2011). In this way, private sector transparency can be defined as the voluntary adoption of policies promoting the transparency of product data and production processes. This is an important distinction—private sector transparency is more than just the disclosure of product data. It requires that product data be aggregated across organizations, industries, sectors, or national boundaries in order to further key policy goals.

In other words, private sector transparency requires engagement with the public sphere. The availability of transparent data about the activities of private and nonprofit organizations broadens the range of organizations and individuals that can potentially hold these organizations to account. This is not a new phenomenon. Instead of engaging in regulatory oversight on their own initiative, which is more resource intensive, government agencies often seek to rely more extensively on third parties to raise the alarm about cases of noncompliance. This decision is referred to in public administration scholarship as a choice between "police patrols" and "fire alarms" (McCubbins & Schwartz, 1984). New technologies, combined with appropriate data transparency, have the potential to increase the scope and scale of "fire alarm" strategies. In other words, smart disclosure strategies are just the beginning of a process that could culminate in the crowdsourcing of regulatory compliance.

1.4.2 Benefits to Private Companies

Many companies in the private sector already realize the benefits that product data disclosure and private sector transparency can bring. These benefits include opportunities for market differentiation: by building a brand, label, tool, or system around product data disclosure, companies can demonstrate that their products are greener, healthier, more local, or more ethical and differentiate them from other products in the market (Howard, 2012; Thaler & Sunstein, 2008). This can be very important in crowded markets where products and services can be very similar, particularly in situations where consumers can observe little from the product packaging itself as to the product's content or quality.

A second, related opportunity for firms is to build brand identification through customer ownership of customer data. By giving consumers ownership of their own data, companies can build tools that increase consumer identification with their products (Thaler & Tucker, 2013). They can learn more about consumer preferences in this way.

For certain companies and entrepreneurs, the disclosure of non-price product information can present new commercial opportunities. One of the core benefits often attributed to online disclosure of data is that other organizations and individuals that use the existing data in innovative ways can emerge. In some cases, this might mean research opportunities, reducing the costs to a company in developing a new product. Releasing product data regarding pharmaceuticals, for example, could allow faster development of more effective medicines or cheaper generic medicines. In other cases, it might mean the development of secondary information filtering service or tool that makes use of the disclosed information. These tools would be targeted at companies as well as consumers who do not have the time or resources to process and filter through large amounts of publicly available data. This is certainly a key motivation of the US government's smart disclosure policies (Sunstein, 2012).

But amid all the optimism regarding the potential of information disclosure via new and emerging technologies as a policy tool, it is easy to forget that many existing regulations already rely on disclosure to deliver policy outcomes. Research on this topic finds that information disclosure alone—even mandatory disclosure—is not always enough to incentivize firms to behave in certain ways. Without additional governance structures, disclosure can easily fail to deliver expected policy outcomes (Kraft, Stephan, & Abel, 2011).

The following section discusses the barriers to making product data public and the governance challenges that these barriers pose. The rest of the volume then makes proposals as to how these challenges might be overcome.

1.5 Barriers to Making Product Data Public

Although current and developing technologies make the disclosure and productive use of private and public data seem more feasible than ever before, data interoperability and disclosure are not solely problems of technology. They are also problems of human interaction. The competing interests of organizations and individuals with a stake in the debate over product data disclosure must somehow be managed.

In promoting smart disclosure, however laudable that goal might be, policymakers are making several key assumptions, including that private sector organizations and public agencies will be willing and able to share their data and that consumers, stakeholder groups, and businesses will be able to use and interpret disclosed data in meaningful and profitable ways.

There are, therefore, several interconnected barriers facing anyone wishing to incentivize the disclosure of product data and promote its meaningful use: problems relating to the cost of disclosure, problems relating to commercial competition and the perceived commercial sensitivity of the data to be disclosed, problems relating to the preservation of privacy, problems relating to data quality and interoperability, and legal barriers to disclosure.

The Cost of Disclosure The first set of dilemmas relates to the organization's decision to disclose data. Data disclosure is not without cost. Organizations may not have access to data in a form that is usable or that makes disclosure viable. Data collection, translation, or reformatting may have to occur. Checking the data for errors also incurs costs, as does managing the disclosure process itself. The key question is—do the benefits of disclosing the data outweigh the costs to an organization? It is also important to ask, what is the necessary and desirable level of information that can be provided at a reasonable cost? The answers to these questions may be different for every organization, but they are also dependent upon what others decide to do. An organization's cost/benefit calculation might change, for example, based on the participation of a critical mass of similar actors or on likely consumer demand for the disclosed data (Ran et al., 2016).

Competition and Commercial Sensitivity "Commercially sensitive information" is a rubric that allows organizations to withhold information of many different kinds. Some companies, particularly those that compete on price, may see their supply chain data as commercially sensitive or as a trade secret and may be wary of revealing it to competitors.

Legal Related to this dilemma is the fact that laws and regulations are often barriers to information disclosure. These legal barriers may be real (enacted to protect consumers or ensure fair competition) or they can be imagined (an excuse not to disclose data).

Data Quality Policymakers must consider not just how to promote the disclosure of more information, but how to improve the quality of the information disclosed. We know that government data often suffers from missing or incomplete information. The complexity of regulatory procedures means that there is considerable scope for errors and omissions. Government inspections of products might be patchy or inaccurately recorded. Compliance reports held by product certifiers may be submitted in hard copy only or in a format (such as PDF) that makes it hard to repurpose the information they contain.

This problem affects the private sphere, too. Large corporations who outsource work to other organizations may not have complete records of the supply chains in which they operate. In fact, it can be to a company's advantage to obfuscate supply chain data, hiding any efficiency advantages (legitimate or questionable) from competitors. Businesses may therefore be unwilling to disclose if they have incomplete records. No matter what the provenance of the data, there is a chance that it could be fraudulent.

Missing, incomplete, or poorly trusted data are problematic and undermine the fundamental goals of smart disclosure, open government, and private sector transparency. The poorer the quality of the data, the higher the cost of utilizing it for other purposes. Private actors or consumers might be uninterested in product data of poor quality. Broker organizations looking to develop consumer tools may well pass on the opportunity to use certain datasets if they calculate that the up-front cost of making the information usable is too high.

Interoperability Data may also be held in a format or structure that allows data from different organizations to be made interoperable. This choice, too, can be deliberate. Disclosing data is really only the first step. Policymakers must also think about how to make the disclosed data interoperable, in order to promote its meaningful use and reuse in ways that promote public value. This is an enormous technical challenge, and any attempt to solve it must rest on cooperation among the various stakeholders. Interoperability requires extensive collaboration between organizations and individuals, something that ultimately rests on establishing trusted relationships among them.

Underpinning these dilemmas is the fact that policymakers have to make some significant predictions about how individual consumers will behave. Smart disclosure policies imply that better data disclosure will lead to an improved information environment for consumers and will impact consumer choice. Producers and retailers are interested in disclosing product data because of the potential to differentiate their products within crowded markets, making them more visible to consumers. Providing trusted information about the origins of a product to consumers can enhance a company's sustainable credentials against its competitors. But this only works if consumer behavior is truly altered by the disclosure of product data.

Therefore, consumer demand for disclosure of privately held product data is central to efforts to encourage private sector transparency. This "demanded disclosure," driven primarily by consumer demand for product data and supplemented by government mandates or companies' attempts to influence markets, is facilitated by new forms of technology that reduce the costs of exerting social pressure on organizations and governments (Sayogo, 2013; van der Laan, 2009).

We do know that consumers' trust in the data provided plays an important role in whether or not they use a particular system (Luna-Reyes et al., 2013, 2014; Sayogo, Zhang, Liu et al., 2014; Sayogo, Zhang, Pardo et al., 2014; Sayogo et al., 2015). Consumers should be protected from fraudulent use of disclosed product data. Any governance system promoting product data disclosure should consider the relationship between collaborative standards for governing product data disclosure and hard law remedies against fraudulent use of product data, certifications, or labels.

Access Versus Privacy An additional key challenge relating to consumer trust is how to protect consumer privacy in an open and accessible system. The difficulties associated with protecting individuals' privacy can form a barrier to disclosure. How should individual and commercial privacy be balanced with appropriate, and broadly applicable, access to information? In order for them to trust the system, consumers should have the right to expect that important personal information will be kept private and the right to be protected from organizations that want to use disclosed information for direct marketing and scams. It is important to avoid disclosing identifiable information—and with multiple organizations disclosing, this requires strong consensus on how to handle and process the data before it is disclosed, as well as an agreement on enforcement mechanisms.

1.6 Concluding Remarks

Technology is facilitating a revolution in the way we access information about markets, lowering dramatically the opportunity costs of learning about the provenance of the things that we make and buy. Meanwhile, policymakers have stated that they wish to encourage the disclosure of product data for a number of reasons, including facilitation of consumer choice, product innovation and research, and creation of more efficient ways to regulate markets.

We argue that in many areas, particularly those relating to sustainability, this process of disclosure can enhance public value by adding value to the public sphere through a fair and transparent process. Introducing new information about production processes into the public sphere, and creating a space whereby the public and other organizations can enter into a dialogue around that information on equal terms with firms and governments, is something that we believe will enhance public value.

This is an ambitious, long-term goal, well suited to the public value approach, which emphasizes long-term, strategic action to align actors toward key policy goals. It cannot be achieved without collaboration among a range of disparate actors—the participatory elements of the public value approach. Collaboration between government agencies, private actors, and consumer advocates is necessary in order to promote the disclosure of product data and to move toward the long-term goals of greater private sector transparency and, ultimately, a more sustainable world.

To answer our research question, we used a variety of methods and approaches. A detailed description of such approaches is included in the methodological appendix to this book, and the results of our inquiry are reported in the following seven chapters of the book.

Chapter 2 introduces the challenges and issues in developing a platform supporting interoperability in sustainable supply chains of food products from the point of view of key stakeholders in the coffee supply chain. The chapter is based on data from a workshop and a series of interviews with stakeholders in the coffee industry. The analysis reveals that to build an interoperable data architecture to support a sustainable supply chain, the five most salient issues/challenges are to build trust in the data, to develop semantic capabilities as well as standards and protocols for data sharing, to design an information policy that balances commercial interest and openness, to establish a collaborative governance model, and to develop a sustainable business model to push forward the vision of the system.

Chapter 3 explores the existing relationships among supply chain participants to better understand current forms of collaboration and the role that trust plays in each supply chain configuration inside the coffee supply chain. The chapter results are grounded on interviews with supply chain participants in Mexico, including small and medium producers, cooperative representatives, large intermediaries, and large corporations. We explore the role and evolution of three different types of trust-producing mechanisms (institutional, calculative, and relational) in the three most common supply chains identified in our field work (large cooperatives, small specialty coffee roasters, and large corporations). Our results suggest that relational trust is more important to facilitate collaboration in both cooperatives and specialty coffee types of supply chains, especially at the beginning of the relationship. Large corporations, like Nestle, rely much more in institutional trust to start collaboration. The main source of such institutional trust is brand reputation and contracts. Relational trust in this type of supply chain is built over time through collaboration. Finally calculative trust plays a role at the start of the collaboration, but loses importance over time in all cases. We conclude the chapter with a reflection on how these types of trust could play a role in building a network of stakeholders around an architecture of shared product information.

Although consumer trust is solicited through the enactment of certification and labeling practices, the rapid growth of certifications and labels has decreased the amount of trust generated from certifications mainly due to the difficulties faced by consumers in observing information behind the labels or certificates. Chapter 4 explores the sufficiency of existing certification and label as part of private regulation for enhancing consumer trust. We evaluate the strengths and weaknesses of the six major coffee certification initiatives by conducting a rigorous document review as well as content analysis of the certification website. Our evaluation found that certification and labeling schemes use different strategies to emphasize the legitimacy and accountability of their practice to assert their trustworthiness, such as openly publishing their standards and principles or getting accreditations from reputable national or international organizations. Our evaluation also demonstrates the complexities of certification and inspection process in the sustainable supply chain. Such complexities challenge the effort to encourage private sector transparency to support interoperable platform such as I-Choose.

Chapter 5 includes one of the key components of our research program, which is a concrete proposal for the creation of a technical architecture and platform to share trusted information about sustainability of products among supply chain stakeholders and consumers. We propose the use of semantic web applications and ontologies to create such an architecture. In this chapter, we outline what we are calling a Certification and Inspection Data Infrastructure Building Block (CIDIBB). CIDIBB involves the interaction of at least three interrelated ontologies. The first of them is a Certification and Inspection Ontology (CerTIN) that defines at a high level of abstraction the main components of any certification system. This high level ontology interconnects more specific ontologies developed for each certification standard. The second ontology (CiTruST) defines the quality of the certification process using the main definitions included in the certification ontology as mechanisms for trust creation. Finally, we include in our proposal the FLO ontology as an example of a local ontology. This ontology was created with the purpose of testing ways in which CerTIN could be mapped to specific certification schemes. Beyond describing the ontologies, the chapter also includes an evaluation of the ontology in their capacity to assess the levels of trustworthiness of different certification schemes and provides examples of ways in which the architecture can be used to create networks of stakeholders around four different business models.

Chapter 6 outlines the privacy, confidentiality, and security issues that are inherent in the design and implementation of IT-enabled platforms such as I-Choose, which was described in detail in Chap. 5. I-Choose enables the implementation of smart data disclosure that requires integration of data from diverse stakeholders in a complex sustainable certified coffee supply chain. Importantly, we discuss these issues from an organizational perspective along three dimensions: ownership, access rights, and data quality. To support the arguments we make in this chapter, we extensively use data from in-depth interviews with the supply chain stakeholders including producers, roasters, exporters, inspectors, certifiers, and consumer advocates. The challenge to protecting the confidentiality and privacy of the data and information lies in developing effective and transparent security policies and protocols that govern the access and integrity to both proprietary and public information. Our findings highlight that these challenges stem from the complexity of the information chain and the value propositions of the various stakeholders in the sustainable coffee supply chain. As a result, addressing these issues necessitates both business practices and governance and not solely technological fixes. Therefore, we propose five management and policy strategies for mitigating the privacy, confidentiality, and security challenges that confront successful implementation of platforms such as I-Choose.

Concluding remarks of most of the first six chapters of the book point out to the importance of finding the proper incentives for supply chain participants, as well as the key role of governance mechanisms. Chapter 7 concentrates on the discussion of incentives and governance. As discussed in Chap.2, it is possible to identify five different configurations in the coffee supply chain in the NAFTA region. As we describe in Chap. 7, stakeholders in each configuration hold different frameworks of reference guiding decisions on quality, means, and ends. We use in the chapter Convention Theory to categorize these reference frameworks as domestic, civic, market, and industrial worlds. The coexistence of those frameworks represents sources of conflict, in addition to power imbalances in each supply chain governance mode, posing specific challenges when introducing a platform like I-Choose into an existing supply chain. Our empirical work shows that in practice, supply chain participants can be characterized by a combination of at least two of such views and that participants in a single supply chain configuration tend to share those views. We also specify the conditions that make different supply chain configurations and set of values more or less amenable to the changes implied in the disclosure of private information that the I-Choose platform requires.

Our concluding chapter draws on the concepts and theories discussed in the book, particularly the concept of public value creation, the conflicts between different types of stakeholders, and the role of trust. The chapter focuses on the practicalities of information disclosure by asking: how must this process be governed? A definition of governance as the process of steering a society toward a set of predefined goals is introduced. It discusses the benefits and difficulties of creating collaborative governance in the context of our project and presents our findings regarding governance from the I-Choose project. It evaluates existing experiments in collaborative governance that aim to extract public value from data disclosure, drawing on examples from multiple countries and cross-border contexts, including the I-Choose project. We find that information disclosure alone is not enough to enhance the public sphere. It must be supported by innovative governance mechanisms that address classic problems such as establishing independence among producing and regulating organizations and creating procedural transparency.

We hope that, through this book, we are contributing to a better theoretical and practical understanding of the different technical, organizational, and policy components needed to create public value through the disclosure of private information inside supply chains.

References

Abolafia, M. Y. (2001). *Making markets: opportunism and restraint on Wall Street* (2. print). Cambridge, MA: Harvard Univ. Press.

Akerlof, G. A. (1970). The market for "lemons": Quality uncertainty and the market mechanism. *The Quarterly Journal of Economics, 84*(3), 488–500.

Baue, B., & Murninghan, M. (2011). The accountability web: Weaving corporate accountability and interactive technology. *Journal of Corporate Citizenship, 41*, 27–49.

Benington, J. (2011). From private choice to public value? In J. Benington & M. H. Moore (Eds.), *Public value: theory and practice* (pp. 31–51). New York: Palgrave Macmillan.

Benington, J., & Moore, M. H. (Eds.). (2011). *Public value: theory and practice*. New York: Palgrave Macmillan.

Boeder, P. (2005). Habermas' heritage: The future of the public sphere in the network society. *First Monday, 10*(9). Retrieved from http://firstmonday.org/ojs/index.php/fm/article/view/1280/1200

Bozeman, B. (2007). *Public values and public interest: counterbalancing economic individualism*. Washington, D.C: Georgetown University Press.

Calhoun, C. J. (Ed.). (1992). *Habermas and the public sphere*. Cambridge, MA: MIT Press.

Castells, M. (Ed.). (2007). *Mobile communication and society: A global perspective: A project of the Annenberg Research Network on international communication*. Cambridge, MA: MIT Press.

Castells, M. (2012). *Networks of outrage and hope: social movements in the internet age*. Cambridge, UK/Malden, MA: Polity Press.

Center for Technology in Government. (2011). *Open government and public value: conceptualizing a portfolio assessment tool*. Retrieved from http://www.ctg.albany.edu/publications/online/pvat/PVAT_ConceptualizingtheTool.pdf

Dunleavy, P. (2002). *Democracy, bureaucracy and public choice: Economic explanations in political science* (Nachdr.). London: Prentice Hall.

Executive Office of the President. (2013). *Smart disclosure and consumer decision making: report of the task force on smart disclosure*. National Science and Technology Council. Retrieved from https://www.whitehouse.gov/sites/default/files/microsites/ostp/report_of_the_task_force_on_smart_disclosure.pdf

Fernback, J., & Thompson, B. (1995). *Computer-mediated communication and the American collectivity: The dimensions of community within cyberspace report as virtual communities: Abort, retry, failure?* Presented at the International Communication Association, Albuquerque, New Mexico. Retrieved from http://www.well.com/user/hlr/texts/VCcivil.html

Fridell, G. (2007). Coffee and the "double movement" (Chapter 4). In *Fair trade coffee: The prospects and pitfalls of market-driven social justice* (Reprinted, pp. 135–172). Toronto: Univ. of Toronto Press.

Fuchs, C., Boersma, K., Albrechtslund, A., & Sandoval, M. (2012). *Internet and surveillance: The challenges of Web 2. 0 and Social Media*. New York: Routledge.

Habermas, J. (1989). *The structural transformation of the public sphere: An inquiry into a category of bourgeois society*. Cambridge, MA: MIT Press.

Howard, A. (2012, April 1). What is smart disclosure? Retrieved from http://radar.oreilly.com/2012/04/what-is-smart-disclosure.html

Jorgensen, T. B., & Bozeman, B. (2007). Public values: An inventory. *Administration & Society, 39*(3), 354–381. http://doi.org/10.1177/0095399707300703.

Kalampokis, E., Tambouris, E., & Tarabanis, K. (2011). A classification scheme for open government data: Towards linking decentralised data. *International Journal of Web Engineering and Technology, 6*(3), 266. http://doi.org/10.1504/IJWET.2011.040725.

Kraft, M. E., Stephan, M., & Abel, T. D. (2011). *Coming clean: Information disclosure and environmental performance*. Cambridge, MA: MIT Press.

Lourenço, R. P. (2015). An analysis of open government portals: A perspective of transparency for accountability. *Government Information Quarterly, 32*(3), 323–332. http://doi.org/10.1016/j.giq.2015.05.006.

Luna-Reyes, L. F., Zhang, J., Roy, R., Andersen, D. F., Whitmore, A., & Andersen, D. L. (2013). Information strategies to support full information product pricing: The role of trust. *Information Polity, 18*(1), 75–91.

Luna-Reyes, L. F., Zhang, J., Whitmore, A., Jarman, H., Picazo-Vela, S., Andersen, D. L., et al. (2014). Full information product pricing: An information strategy for harnessing consumer choice to create a more sustainable world. *Communications of the Association for Information Systems, 34*(1), 637–654.

McCubbins, M. D., & Schwartz, T. (1984). Congressional oversight overlooked: Police patrols versus fire alarms. *American Journal of Political Science, 28*(1), 165. http://doi.org/10.2307/2110792.

Michalos, A. C. (2008). *Trade barriers to the public good: Free trade and environmental protection*. Montreal: McGill-Queen's University Press.

Moore, M. H. (1995). *Creating public value: Strategic management in government*. Cambridge, MA: Harvard Univ. Press.

Moore, M. H. (2013). *Recognizing public value*. Cambridge, MA: Harvard University Press. Retrieved from http://www.hup.harvard.edu/catalog.php?isbn=9780674066953.

Nescafe. (n.d.). The Rainforest Alliance – working together for sustainable agriculture. Retrieved from http://www.nescafe.com/rainforest_alliance_sus_en_com.axcms

Niskanen, W. A. (1994). *Bureaucracy and public economics*. Aldershot, Hants, England; Brookfield, VT: E. Elgar.

Olson, M. (1965). *The logic of collective action: Public goods and the theory of groups*. Cambridge, MA: Harvard Univ. Press.

Paige, J. M. (1997). *Coffee and power: revolution and the rise of democracy in Central America* (1. Harvard Univ. Press paperback ed., 3. printing). Cambridge, MA: Harvard University Press.

Rainforest Alliance. (2015). *Our global impact: Agriculture*. Retrieved from http://www.rainforest-alliance.org/work/impact/map/agriculture

Rainforest Alliance. (n.d.). *Our work with Nescafé*. Retrieved from http://www.rainforest-alliance.org/about/company-commitments/nescafe

Ran, W., Jarman, H., Luna-Reyes, L. F., Zhang, J., Andersen, D. L., Tayi, G. K., et al. (2016). Supply-chain transparency and governance systems: Market penetration of the I-choose system.

In J. Zhang, L. F. Luna-Reyes, T. A. Pardo, & D. S. Sayogo (Eds.), *Information, models, and sustainability: Policy informatics in the age of big data and open government*. Basel, Switzerland: Springer.

Rheingold, H. (2008). Using participatory media and public voice to encourage civic engagement. In W. Lance Bennett (Ed.), *Civic life online: Learning how digital media can engage youth* (pp. 97–118). Cambridge, MA: The MIT Press.

Rice, R. A., Ward, J. R., Smithsonian Migratory Bird Center, & Natural Resources Defense Council. (1996). *Coffee, conservation, and commerce in the Western Hemisphere: How individuals and institutions can promote ecologically sound farming and forest management in northern Latin America*. Washington, D.C.; New York: Smithsonian Migratory Bird Center; Natural Resources Defense Council.

RussiaToday. (2013, August 21). NSA collected thousands of US internet communications "with no terror connection" [News]. Retrieved from http://on.rt.com/icwoby

SAGARPA. (2012). *Impactos del Café en México*. Secretaría de Agricultura, Ganadería, Desarrollo Rural, Pesca y Alimentación. México, D. F.

Sayogo, D. (2013). *Incentives to disclose: Private sector transparency to support smart disclosure policy (Unpublished doctoral dissertation)*. University at Albany, SUNY, Albany, NY.

Sayogo, D. S., Zhang, J., Liu, H., Picazo-Vela, S., & Luna-Reyes, L. (2014). *Examining trust as key drivers in smart disclosure for sustainable consumption: The case of I-choose*, In Proceedings of the 15th Annual International Digital Government Research Conference on Open Innovations and Sustainable Development in Government: Experiences from around the World, Digital Government Society of North America (pp. 137–146). Aguascalientes, Mexico: ACM.

Sayogo, D. S., Zhang, J., Pardo, T. A., Tayi, G. K., Hrdinova, J., Andersen, D. F., et al. (2014). Going beyond open data: Challenges and motivations for smart disclosure in ethical consumption. *Journal of Theoretical and Applied Electronic Commerce Research, 9*(2), 3–4. http://doi.org/10.4067/S0718-18762014000200002.

Sayogo, D. S., Zhang, J., Luna-Reyes, L., Jarman, H., Tayi, G., Andersen, D. L., et al. (2015). Challenges and requirements for developing data architecture supporting integration of sustainable supply chains. *Information Technology and Management, 16*(1), 5–18. http://doi.org/10.1007/s10799-014-0203-3.

Sayogo, D. S., Ran, W., Tayi, G. K., Luciano, J. S., Luna-Reyes, L. F., Depaula, D., et al. (Forthcoming). Ontological modeling of certification and inspection process to support smart disclosure of product information. *International Journal of Public Administration in the Digital Age*.

Schattschneider, E. E. (1935). *Politics, pressures and the tariff: A study of free private enterprise in pressure politics, as shown in 1929–1930 revision of the tariff*. New York: Prentice-Hall.

Smith, T. W. (2015). *Trends in national spending priorities, 1973–2014* (General Social Survey). Chicago: NORC at the University of Chicago. Retrieved from http://www.norc.org/PDFs/GSS%20Reports/GSS_Trends%20in%20Spending_1973-2014.pdf

Stoker, G. (2006). Public value management: A new narrative for networked governance? *The American Review ofPublic Administration, 36*(1), 41–57. http://doi.org/10.1177/0275074005282583.

Sunstein, C. (2012, March 30). Informing consumers through smart disclosure. Retrieved from http://www.whitehouse.gov/blog/2012/03/30/informing-consumers-through-smart-disclosure

Thaler, R. H., & Sunstein, C. R. (2008). *Nudge: Improving decisions about health, wealth, and happiness*. New Haven: Yale University Press.

Thaler, R. H., & Tucker, W. (2013, February). Smarter information, smarter consumers. *Harvard Business Review*. Retrieved from http://hbr.org/2013/01/smarter-information-smarter-consumers

USDA Foreign Agriculture Service. (2011). *Coffee consumption growing and producer inventories building* (Required Report – public distribution No. MX1039). Mexico City: US Department of Agriculture.

van der Laan, S. (2009). The role of theory in explaining motivation for corporate social disclosures: Voluntary disclosures vs "solicited" disclosures. *Australasian Accounting, Business and Finance Journal, 3*(4). Retrieved from http://ro.uow.edu.au/aabfj/vol3/iss4/2

Wankhade, L., & Dabade, B. M. (2006). Analysis of quality uncertainty due to information asymmetry. *International Journal of Quality & Reliability Management, 23*(2), 230–241. http://doi.org/10.1108/02656710610640961.

Wilson, W. (1886). The study of administration. *Political Science Quarterly, 2*(2), 197–222.

Zuiderwijk, A., Jeffery, K., & Janssen, M. (2012). The potential of metadata for linked open data and its value for users and publishers. *Journal of E-Democracy and Open Government, 4*(2), 222–244.

Chapter 2
Challenges to Developing Interoperable Data Architecture to Support Sustainable Consumption and Sustainable Supply Chains

Jing Zhang, Djoko S. Sayogo, Giri K. Tayi, and Luis F. Luna-Reyes

Abstract This chapter focuses on the identification of key challenges to building a data architecture to improve sustainability in supply chains as well as providing consumers with better information for decision support. The chapter builds on the trends of sustainable consumption and sustainable supply chain management and incorporates the views of key stakeholders in the coffee supply chain that we interviewed. Key challenges relate to accuracy and credibility of data in the system, to the availability of technical expertise and infrastructure across the supply chain, as well as with legal aspects related to data ownership, privacy, and confidentiality. Finally, finding appropriate ways of funding the architecture constitutes another important challenge.

Keywords Information asymmetry • Supply chain transparency • Sustainable consumption • Sustainable supply chains • Smart disclosure

---

J. Zhang (✉)
Graduate School of Management, Clark University, Worcester, MA, USA
e-mail: jizhang@clarku.edu

D.S. Sayogo
Department of Economics and Business, University of Muhammadiyah at Malang, Malang, Indonesia

Center for Technology in Government, University at Albany
187 Wolf Rd., Suite 301, Albany NY 12205, USA
e-mail: dsayogo@ctg.albany.edu

G.K. Tayi
Department of Information Technology Management, University at Albany, Albany, NY, USA
e-mail: gtayi@albany.edu

L.F. Luna-Reyes
Department of Informatics, University at Albany, Albany, NY, USA
e-mail: lluna-reyes@albany.edu

H. Jarman, L.F. Luna-Reyes (eds.), *Private Data and Public Value*,
Public Administration and Information Technology 26,
DOI 10.1007/978-3-319-27823-0_2

2.1 Introduction

The issue of sustainable development is not new; the seemingly contradictory goals of economic development and environmental preservation have been encountered by mankind throughout its history. The well-known argument in *The Tragedy of the Commons* (Hardin, 1968) is based on a nearly 200-year-old observation of shepherds' treatment of common grazing pastures by an English economist, William Forster Lloyd. But the history of a concerted movement on the part of governments toward sustainable development and consumption is relatively recent, marking its official start with the worldwide adoption of *Agenda 21* at the 1992 Earth Summit in Rio de Janeiro. The 178 governments that voted to adopt *Agenda 21* recognized the importance of sustainable development and committed themselves to the promotion of more sustainable consumption patterns.

The World Commission on Environment and Development (WCED) defines sustainability as "using resources to meet the need of the present without compromising the ability of future generations" (Linton, Klassen, & Jayaraman, 2007). A mainstream approach to promoting sustainable consumption consists of government interventions to correct prices and provide regulatory frameworks to nudge producers into becoming more eco-efficient (Seyfang, 2005). Unfortunately, this approach has achieved only limited success, at least in part because of barriers to the dissemination of sufficient product information among all actors along the supply chain. In the vast majority of traditional supply chains, consumers have only limited access to information about products' environmental and social impacts and are thus forced to rely only on price and information printed on the product package to help them with their purchasing decision. This lack of information creates difficulties for consumers who are interested in assessing and understanding the implications of their consumption choices (Seyfang, 2005). But the lack of integrated information about environmental impacts in the sustainable supply chain does not create problems only for the consumer, but it also has the potential to disrupt the entire system. Importers, for example, need assurance of their supplier's compliance with independent or voluntary codes of conduct or public standards (Vachon & Klassen, 2007). As we have seen in many recent examples, such as the Rana Plaza factory collapse,[1] the lack of knowledge about production practices or safety standards employed by overseas subsidiaries can have serious negative impacts on the reputation of major brands, subjecting them to economic and reputational losses.

As a result, governments begun to adopt a new approach to promoting sustainability that focuses on providing access to information to increase market transparency and efficiency, which in turn can lead to sustainable production and consumption. This new approach is based on the assumption that better informed consumers will make purchasing decisions that will provide incentives for the whole

[1] http://www.theguardian.com/sustainable-business/2015/jun/10/rana-plaza-fund-reaches-target-compensate-victims

supply chain to move toward a more sustainable production model. The smart disclosure initiative of the Obama Administration is one example of such effort to promote innovations that help consumers use their social and environmental values to guide their marketplace decisions (Howard, 2012; Thaler & Sunstein, 2008).

The new approach partially stems from recent technological developments that have the potential to improve transparency of the supply chain processes by streamlining the information flow from the producer to a consumer. Such streamlined information flow could reduce information asymmetry in the supply chain (Jarman et al., 2011), which is generally characterized by consumers not having enough information to verify supplier/buyer behavior (Eisenhardt, 1989; Fama & Jensen, 1983) or by being unable to accurately evaluate information quality and breadth (coverage) (Akerlof, 1970; Mishra, Heide, & Cort, 1998). Reducing barriers to information flow would benefit not only consumers but also other actors in the supply chain, such as producers, retailers, distributors, and certifiers. Well-known benefits include, for example, reductions of costs of coordination and increased flexibility (Clemons & Row, 1993; Malhotra, Gosain, & Sawy, 2005; Wang, Tai, & Wei, 2006). Less-studied potential benefits include the use of shared information to produce knowledge that can be used as marketing intelligence, allowing members of the supply chain to find new markets for their products (Malhotra et al., 2005). However, governance plays a key role in the distribution of benefits, which may in turn become one of the main barriers for sustainable coordination and information sharing (Clemons & Row, 1993; Johnston & Vitale, 1988).

As described in Chap. 1, the I-Choose project aimed at reducing information asymmetry by developing an interoperable data architecture that would support integration of information along the sustainable supply chain for coffee produced and consumed in the NAFTA region. We argue that by allowing for a more direct connection between the consumer and the producer, such data architecture would enable innovations and changes conducive to the development of a more sustainable production and consumption environment. This chapter focuses on the challenges to developing an interoperable data infrastructure for the sustainable coffee supply chain from the standpoint of key stakeholders as well as necessary conditions that would make such development possible. Both the challenges and the system conditions were identified through our workshops and interviews with various actors involved in the sustainable coffee supply chain.

This chapter is organized into five sections, including the foregoing introduction. Section two provides a literature review focused on the role of information in sustainable consumption, smart disclosure, and sustainable supply chain management. The third section outlines the key stakeholders of the sustainable coffee supply chain, their interconnectedness, and the flow of data that is relevant to assessing sustainable practices. The fourth section identifies challenges to developing an interoperable data architecture to share information from the perspective of key stakeholders. The concluding section then focuses on the conditions and system requirements that are needed for the development of integrated data architecture to be possible.

2.2 Literature Review

Over the last few decades, the role of information in supporting sustainable consumption has been studied by various disciplines, each approaching the topic from a different standpoint. In the first part of this section, we focus on reviewing literature that identifies information asymmetry as a major barrier to developing the necessary market conditions for further promotion of sustainable consumption and the open data movement as an opportunity to address this problem. In the second part, we review the literature focusing on promotion of sustainability through a complete supply chain approach and how information integration across the supply chain could help decrease its information asymmetry (Seyfang, 2005).

2.2.1 Sustainable Consumption, Information Barriers, and Open Data

Although there is still no consensus on the exact definition of sustainable consumption (Mont & Plepys, 2008), the literature generally identifies two mainstream approaches to promoting sustainable consumption: (1) creating eco-efficiencies and "greening" the production processes and (2) changing consumer consumption levels and patterns (Fuchs & Lorek, 2005; Mont & Plepys, 2008; Seyfang, 2005). Although the first approach, generally characterized by increasing production efficiencies and governmental regulations aimed at lowering environmental impact of production, has had a wide acceptance among governments, the second approach is regarded as having a potentially higher impact due to the potential for lower consumption levels overall (Fuchs & Lorek, 2005). The second approach attempts to improve sustainability by setting up policies to influence consumers' consumption behavior (Seyfang, 2005) and relies on market forces and signals from consumers regarding consumption habits. In this way, increasing demand for "sustainable" products will eventually transform production practices and supply chain processes. This mainstream approach, highly dependent on consumer behavior, is thus prone to failure if barriers to information flow are present (Seyfang, 2005). Having all the necessary information, a consumer could scrutinize the supply chain, assessing how well practices and processes in it match her personal values to make a sustainable purchasing decision (Jarman et al., 2011). Unfortunately, in majority of supply chains, consumers and other actors make decisions based upon limited information.

Of course the degree of information asymmetry depends on the type of product attributes one is searching for (Darby & Karni, 1973; Nelson, 1970). Nelson (1970), for instance, distinguished between search and experience attributes. A search attribute (e.g., price) can be known before the purchase, and consumers have the ability to search for it and can let it influence their purchasing decision. Experience attributes (e.g., flavor) are only known after the consumer experiences the product (Nelson, 1970), and thus a consumer cannot use them to help him make the initial purchasing decision. However, even though experience attributes are not knowable prior to the purchasing decision, the consumer can use this information about his experience when making purchasing decisions in the future. Credence attributes (e.g., the use of organic practices), on the other hand, are not detectable to the consumer even after experiencing the product, and he is thus forced to rely on third-party judgment or certification (Darby & Karni, 1973).

The limited availability and unequal access to product information limit the ability of consumers to understand the environmental and social implications of their consumption decisions. In this way, eliminating or reducing information asymmetry is necessary for supporting sustainable consumption (Senge et al. 2008; Goleman 2010). Efforts to reduce asymmetry of information related to environmental and social impacts have led to the proliferation of third-party certifications and labels (Jahn, Schramm, & Spiller, 2005), which attempt to help consumers differentiate between organizations and products in respect to their environmental and social practices. Unfortunately, these efforts are challenged by the ambiguity of what the various labels represent and the varied degrees of rigor behind each certification. Due to the diversity of certifications and labels, consumers might very well be wary of the meaning or credibility of information provided in the labeling scheme. Critical voices have also questioned whether the voluntary measures provided through labeling and limited product information were not a corporate "greenwash" or "bluewash" (Clapp, 1998; Fuchs & Lorek, 2005). Based on these limitations of third-party certifications, Jahn et al. (2005) added a fourth dimension to product description, "Potemkin attributes." Potemkin attributes are visible only after a close examination of the internal processes used in producing and handling of a given product and are made available by tracing the provenance of information along the supply chain (Jahn et al., 2005).

The context described above constitutes an opportunity for open data efforts and for promoting innovation among a diverse set of individuals and organizations, including producers, supply chain operators, certifiers, government agencies, NGOs, and information aggregators that provide analyzed information to consumers. Since 2009, US government and governments around the world have developed policy initiatives for promoting disclosure of information held by both public and private entities through open government and smart disclosure efforts. The goal of smart disclosure is to foster the creation of products and tools that help consumers make important marketplace decision. Although federal government has historically disclosed consumer information, the rise of Web 2.0 and Internet technology enabled governments to use information disclosure as a policy approach in areas such as health, education, energy, finance, and public safety. The highlighted benefits of smart disclosure include enabling consumer decision-making in complex market conditions and improving economy by enhancing market transparency and efficiency. In order to achieve these benefits, it is recognized that disclosure of information held by public entities is not enough as vast amount of information that is relevant to consumer choices are held by private entities. To provide consumers with access to product information, companies involved in production and distribution of products need to be encouraged to release data that is of high quality and in machine readable format.

2.2.2 Information Technology and Sustainable Supply Chain Management

Wider interest in the issue of sustainability from the standpoint of industry begun with the recognition of companies as important stewards in addressing the challenges of sustainable development in the 1990s (Angell & Klassen, 1999; Matos & Hall, 2007). This recognition in turn created immense external pressure from government, the public, and nongovernmental organization forcing companies to integrate sustainability into their practices (Linton et al., 2007; Sarkis, Zhu, & Lai, 2011; Vachon & Klassen, 2007; Zhu & Sarkis, 2004).

The initial focus of companies was on integrating sustainability into their internal operations and reducing the adverse impact of their own organization (Hsu, Tan, Zailani, & Jayaraman, 2013). This strategy was soon deemed inadequate because of the interconnectedness of supply chain partners (Hsu et al., 2013). In reaction, companies started to explore management approaches that consider the environmental impact of the entire supply chain, from production, consumption, and customer service to post-consumption (Hsu et al.; Linton et al., 2007; Matos & Hall, 2007; Vachon & Klassen, 2007). Integrating sustainability management into the entire supply chain is considered to optimize the operation of the company (Linton et al., 2007) and minimize the risks of business operation and global competition (Beamon, 1999; Hsu et al.).

However, in order to manage sustainability across the whole supply chain, it is necessary to share information about the sustainable practices and capabilities of all supply chain partners and to collaborate with each other to address various issues (Vachon & Klassen, 2007). This becomes especially important when consumer interest in company's sustainability practices is heightened, particularly in the food and agriculture industry (Collins, Steg, & Koning, 2007; Locke, Kochan, Romis, & Qin, 2007; Locke & Romis, 2007; Opara, 2003; Wilson & Clarke, 1998). Thus information technology that enables information sharing and integration can play an important role in supporting sustainable supply chain as a whole.

This promise of information technologies, however, is yet to materialize, as information-driven supply systems and infrastructure are in the beginning stages of design and development (Steinfield, Markus, & Wigand, 2011). In practice, many barriers exist preventing full integration of information across the supply chain. There is as much, if not more, incentive for hoarding or manipulating access to information, as there is for sharing information (Mishra et al., 1998). This opens the door to questions concerning stakeholders' motivations to share information, distribution of benefits, and shifts of bargaining power in supply chains (Clemons & Row, 1993; Johnston & Vitale, 1988). Information systems that would dramatically reduce information asymmetries in supply chains would undoubtedly alter many important attributes of supply chains such as governance, incentives to share information and other resources, and motivations to create and participate in long-term relationships.

In addition, supply chain actors must consider the cost and potential return as developing and maintaining sustainable supply chain management infrastructure will not be free. Creating and sustaining sharing capabilities will impose additional costs on producers, retailers, and everyone in between. On the other hand, current IS research has shown that information sharing promotes efficiencies and cost reductions (Malhotra et al., 2005) and information sharing may create value through marketing intelligence (Malhotra et al., 2005). Yet, the positive impacts of such investments remain difficult to quantify and thus justify within the corporate structure (Wolf, 2011). Likewise, competition and market conditions make the distribution of such benefits inequitable and uncertain along the supply chain, further reducing possible incentive for information sharing (Clemons & Row, 1993; Han, Chang, & Hahn, 2011; Johnston & Vitale, 1988).

Summarizing the literature review on sustainable consumption and sustainable supply chain, it is reasonable to argue that developing information integration capability through an interoperable data architecture is a viable solution to some of the major roadblocks toward an expanded market for sustainable products. The development of such interoperable architecture, however, requires policymaker, managers, and developers to deal with a few key challenges facing the various players along the supply chain.

2.3 The Primary Data Producers and Information Flow in Sustainable Coffee Supply Chain

The coffee supply chain in the NAFTA region consists of multitude of actors with different roles and different responsibilities in respect to data production, data maintenance, and data ownership. Identification of these main actors is key to understanding the data environment in the supply chain and to identifying challenges to integrated information sharing. This section presents our findings about the primary data producers and information flows in the sustainable coffee supply chain based on data collected in workshops and 44 semi-structured interviews conducted throughout the I-Choose project.

The workshop participants from the I-Choose project identified seven primary stakeholders in the sustainable coffee supply chain: consumers, producers, certifiers, retailers/roasters, distributors, cooperatives, and consumer associations (Sayogo et al., 2012). Individual interviews conducted as part of the I-Choose project with various stakeholders in the supply chain provided a more accurate account of the involvement of key stakeholders in the creation of data relevant to assessing sustainable practices, the direction of the data flow, and the challenges of data sharing and reuse.

In an ideal supply chain, the relationship between producers, retailers, and consumers is rather linear in respect to product flow which is not necessarily accompanied by a related production information flow. In the case of a sustainable coffee

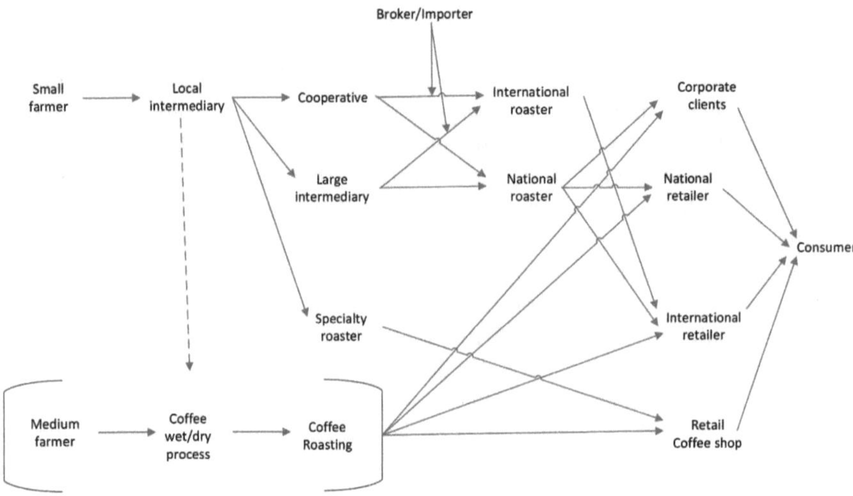

Fig. 2.1 Main stakeholders in sustainable certified coffee supply chain

supply chain, the actors and their relationships are more complex. In Fig. 2.1, we depict several tiers of supply chain actors and the flow of product among them. This figure does not depict flow of information that accompanies products throughout the sustainable coffee supply chain; the flow of information is described in latter paragraphs in this section.

The sustainable coffee supply chain in Mexico includes two different kinds of coffee producers, small and medium farmers. There are no large farmers in Mexico because of land regulations that started after the Mexican revolution in the early twentieth century. Medium farmers tend to integrate the value chain from production to roasting, and most of them have their own brands that are sold in national and international markets. Some medium farmers who are interested in exporting have some coffee certifications such as UTZ or the Rainforest Alliance. Small farmers, on the other hand, only have access to the market through a local intermediary who may have access to larger intermediaries in the supply chain.

Larger intermediaries in the supply chain include cooperatives, specialty roasters, and other larger entities, each of them buying product from different kinds of local intermediaries and selling to a number of various types of clients. In the case of cooperatives, for example, local intermediaries are cooperative members at the local level. In the case of specialty roasters, the local intermediary is usually a well-known certified coffee taster who builds relationships with producers to improve their production methods, so they can provide the specialty roaster with high-quality coffee to be sold directly to the end consumer in small coffee shops. Finally, other large intermediaries gather coffee to be sold to either national or international rosters and are usually in charge of the drying process, getting the coffee beans ready for roasting. Large intermediaries in the sustainable supply chain that we interviewed were certified by 4C and belonged to the Nestlé supply chain.

The interview findings and secondary data analysis indicate two major types of information flow along the supply chains: (a) trading information and (b) certification information. In general, the information collection process for certification purposes begins when[2] an applicant submits an application package and documentation to a certification body. This package is evaluated to determine the eligibility and the scope of audit for the applicant, and if deemed eligible, an authorized auditor conducts an audit on behalf of the certification body. The audit process involves several information collection processes: initial meeting, document review, interviews, physical site visit, and closing meeting. Following the audit, an audit report is sent to the certification body for further evaluation and subsequently for a certification decision. Even though the information is collected by a certification body, the ownership of all data collected during a certification process remains with the applicant, generally producers or roasters. The certification body acts only as an information custodian and stores the information in their database.

Information about products is also collected continuously during the trading process between producers and roasters/importers[3] through trading documents such as invoice, bills of lading, or other documents, to ensure that the traded products are certified as sustainably produced. Each document for sustainable trading must be identified with particular certification body that the producers or roasters are registered with, usually in the form of a certification ID. In addition to the trading documents, some certification bodies require the roasters to submit quarterly reports of their sustainably certified coffee sales.

Given the complexities of the sustainable coffee supply chain and the number of stakeholders involved, it is not difficult to discern the difficulties of integrating information among these actors. In the next section, we describe five categories of challenges identified during our project as they apply to three primary data producers in the supply chain—producers (small and medium farmers), roasters, and certifiers. More detailed descriptions of these challenges are provided in Sayogo (Sayogo, 2013; Sayogo et al. 2014, 2015).

2.4 Challenges to Opening Data to Support Interoperable Data Platforms

The literature on information integration and challenges associated with such endeavors is numerous and spans a number of disciplines from science to technology, business, and medicine. Each information integration effort is unique, however, and as such entails unique challenges based on the complexity of the environment,

[2] Detail description of data collection process can be found in documentations published by third-party certification bodies. For example, for FLO refer to http://www.flocert.net/fairtrade-services/fairtrade-certification/how-it-works/

[3] Some roasters also act as retailers by directly selling their product to end consumers, some roasters sell their product to retailers, and some roaster do both.

the scope of the information landscape, the number of actors involved, and many other factors. The sustainable coffee supply chain is no different, creating a set of unique challenges for the various stakeholders involved in data production along the chain. Throughout our interviews and workshops, we have identified five main challenges to making coffee certification information more readily available to the public that producers, roasters, and third-party certifiers face as the primary data producers, data owners, and/or data stewards.

2.4.1 Data Challenges: Collection, Accuracy, and Credibility

Interoperable data platforms, such as I-Choose, depend for its usefulness and accuracy of its recommendations on unfettered access to data from data producers and data stewards. The first step to making data from the coffee certification process available is to ensure that the necessary data is collected and is accurate and credible. Our interview findings indicate that the main data producers and data stewards—coffee producers, roasters, and third-party certifiers—face a number of challenges when attempting to collect accurate and credible data.

Coffee Producers

As we have seen in Fig. 2.1, producer in a sustainable certified coffee supply chain is usually a cooperative comprised of a number of small farmers or a medium-sized plantation owner. Certification and inspection data for certified sustainable coffee consists of many pieces of data collected from small farmers by either their cooperative or a local intermediary. Farmers represent the smallest unit of data source, and ensuring continuous flow of data from them is crucial for the certified coffee supply chain. Our interviews with various stakeholders indicated that requesting farmers to maintain consistent documentation of their production processes is the most difficult challenge. Typically, farmers do not understand the value of maintaining documentation, see documentation as a waste of their time, and are reluctant to record information about their products if the documentation process is complicated. As a consequence, cooperatives frequently have to assume additional costs to ensure data is collected. Cooperatives often distribute predefined and easy-to-fill forms to farmers or assign staff, usually the internal control, to solicit data from farmers through interviews.

Additionally, large intermediaries such as cooperatives or specialty roasters have to contend with ensuring accuracy and credibility of data in situations where small farmers, motivated by geographical location and financial issues, sell their coffee to local intermediaries who in turn sell it to the large intermediaries. Our interviewees stated that these local intermediaries, especially those who are not part of a cooperative, occasionally fabricate information about the certification of their product in order to quickly fulfill orders. Most often these local intermediaries record and report noncertified products as certified.

Roasters

Roasters' primary challenge is ensuring access to production data and credibility of the data they receive. Roasters generally procure their coffee from traders or importers. Our interviews indicated that some importers hesitate to reveal their information or source of information due to two reasons. First, some traders in a sense "cheapen[ing] the spirit of fair trading" by manipulating prices during negotiations with producers. These traders thus do not wish to disclose their source of product and producer information. Second, hoarding or hiding information is seen by some traders as a competitive advantage and strategy to mitigate threat of substitution. As a result, roasters are often faced with incomplete and inaccurate data regarding the origin of the product they purchase, which in turn affects their ability to provide accurate and complete data for their roasted product.

Third-Party Certifiers

Third-party certifiers face similar challenges as those faced by data producers. Our interviewee from Control Union, a globally recognized certification agency, indicated that inspectors who conduct audits in the field often encounter incomplete or unavailable information for their audit. Since audits are conducted only once a year, the information gaps hinder the ability of inspectors to understand what happens in the period between audits. In addition, despite the existence of criteria that are in turn governed by standards, inspectors and evaluators still need to exercise their own judgment when applying the criteria in the field. For example, the third-level indicators for criteria 3.2.22 of Fairtrade (FLO)—training members on appropriate use of fertilizers—are (a) at least 50 % members have been trained and (b) content of training was sufficient. The inspectors need to use their judgment based on their expertise and experience to measure the sufficiency of the training content. When consistent and reliable information is lacking, the audit reports that are at the core of certification and inspection data are prone to human judgment biases. Without adequate information, auditors must use their experience in deciding the extent to which the applicant conforms to the given criteria and standard.

2.4.2 *Technology Capability: Technical Expertise, Hardware, and Communication*

The second large challenge facing producers, roasters, and third-party certification bodies is limited technological capabilities including limited technical expertise of people on the ground, limited access to technology (hardware), and lack of access to reliable communication infrastructure.

Coffee Producers
Limited access to technology and limited technical knowledge represent a major challenge for both kinds of coffee producers—small farmers and local cooperatives. The lack of technical expertise and access to technology is especially severe at the small farmer level. Our interviews with small farmers point to the challenge of accessing technology due to the geographical remoteness of their location. For instance, when trying to sell organic coffee, such as US Department of Agriculture (USDA) organic, small farmers have difficulties with publishing their organic certificate, which requires information technology as well as a reliable internet connection to access the certifier information system. Both tend to be in short supply in some of the remote areas of Mexican coffee-producing states.

Limited access to technology at points of production also represents a challenge for local cooperatives that are forced to assign additional staff in locations where technology is accessible to facilitate communication with traders (roasters and importers). An additional challenge facing local cooperatives is acquiring the necessary technical knowledge that would be required to manage an implementation of a system that would allow them to digitally integrate data from individual small farmers where they have to adopt computerized record keeping.

Roasters and Third-Party Certifiers
Interviewees from roasters and third-party certifications asserted that they are sometimes limited by their lack of technical expertise despite their willingness to disclose their data. For instance, our roasters/importers' interviewees abandoned their efforts to publish coffee contract documents online due to the lack of sufficient technical expertise that would be needed to build a robustly protected and highly functional database system. Our interviewee from Fair for Life also stated that their organization limits the publication of certification ratings and indicators on their website due to technological capability concerns. The interviewee indicated that they have to use the help from external programmers, which is an additional cost. While larger organizations such

(continued)

as FLO might have the resources to build and maintain a system that allows public to access information about certification, smaller certification systems with smaller target markets simply do not possess the financial resources needed. Although large organizations like FLO do make their systems available to other participants in the system, the licensing costs are high for small and medium roasters and third-party certifiers. The diversity of systems developed by small and medium roasters, besides their technical limitations, poses the additional challenge of integration of information.

2.4.3 *Challenges to Third-Party Certifier: Data Ownership and Conflict of Disclosure Policy*

Majority of data related to product inspection and certification as well as to sustainable trading is in the stewardship of third-party certification bodies that collect and store such data. Opening the data available in third-party certification databases could provide substantial benefits for promoting sustainable consumption practices by lowering information asymmetry in the supply chain through innovative technology solutions that would make such information accessible in an easily understood format. However, third-party certifiers face two interrelated challenges to publishing their data: data ownership and conflict with applicant's disclosure policy.

Data Ownership

Even though third-party certification bodies are responsible for collection and maintenance of certification data, the data ownership remains with the applicant, usually the producer or roaster, not the certification body itself. Thus, any release, access to, or publishing of this data requires consent from the applicants, which is often difficult to obtain, not only because producers need to be willing to disclose their information but also because producers—as we described before—are usually large cooperatives or federations of cooperatives with thousands of members. In such situations, reaching an agreement to disclose data is a very challenging task. Moreover, most certifications are designed with the assumption of data confidentiality as part of their core strategy to attract new producers. Even those that promote a more open system face this challenge on a regular basis. Interviewees from Fair for Life certification—a certification system with an open policy—indicated that the option for data owners to publish certification results is voluntary, and applicants can opt out from this requirement. Consequently, certifiers are limited in the amount of data they can publish.

Conflict of Disclosure Policy

As we briefly commented in the previous paragraph, the disclosure policy of the certification body often conflicts with the disclosure policy of the applicant with regard to publishing certification data. According to an interviewee from CERTIMEX, one of the first things that auditors do when they start working with an applicant is to sign a legally binding document that acknowledges that the ownership of all certification-related data remains with the applicant. Applicants, in the meantime, have their own information policies governing the release of their data, which generally operate on full consensus of all cooperative members. Given the number of farmers and their limited incentive for releasing such data, acquiring a permission for release is quite difficult and time consuming. Other types of applicants, such as medium farmers, take into consideration their own disclosure policy before deciding whether to publish their certification data, and they are especially concerned about the impact to their brand and reputation.

2.4.4 *Information Policy: Confidentiality, Commercial Privacy, and Economic Value of Information*

While disclosure of information has the potential to greatly benefit the public as well as the entity disclosing such information, disclosure also carries risks associated with safeguarding information that might hurt the organization both in the short and in the long term. Our interviews indicated that many companies lack policies that would clearly identify which and how much information should and should not be disclosed due to concerns related to privacy, confidentiality, and economic competitiveness. Interviewees from mission-driven companies pointed out that a major barrier to opening their data is making a decision about how to balance disclosing enough information to create value with the need not to violate disclosure restrictions. Interviewees from corporate entities within the supply chain showed willingness to open their data if doing so adds value to the organization and restrictions related to the confidentiality and economic value of information can be implemented easily. Some information is closely related to competitive advantage and disclosing it potentially endangers the organization's market share.

There is also the issue of ensuring commercial privacy. As alluded to in the paragraph above, disclosing certification results might compromise the reputation and competitive advantage of the applicants. The problem is not necessarily new because supply chains have a long tradition of sharing information to reduce costs and improve flexibility (Malhotra et al., 2005; Clemons & Row, 1993). However, it has also been identified in the literature that the distribution of both costs and benefits can be one of the main barriers to continued information disclosure inside the supply chain. The issue becomes more salient when considering not only these private exchanges but also public disclosure of information.

2.4.5 *Financial Costs of Digital Data/Information Disclosure*

As discussed above, the costs and challenges associated with information collection are tremendous. But the costs associated with information disclosure are not limited just to the point of collection but also involve costs associated with disclosing such information in a digital, machine readable format, as well as the continuous process of updating and maintaining the shared resource. Such information disclosure can be very costly to all actors involved with data production and maintenance.

The interviewees indicated that indirect costs of certification can be very expensive even if mechanisms exist to mitigate direct costs. For instance, third-party certification generally requires producers to maintain records and documentation to support the certification and/or traceability efforts. Maintaining records and documentation represents a major cost for producers due to the challenges discussed in sections above. Publishing their data online is costly for both roasters and third-party certification bodies as well. For roasters, additional work is required to transfer information from an offline format to online or to public domain, which increases expenses. For third-party certification bodies, the information can be extensive; for example, audit results might consist of 30 pages with 10 or 20 control points for each category, making open data costly for them.

As the foregoing discussion indicates, the challenges to creating an interoperable data sharing platform are numerous. As with any information sharing effort, it is not just technical limitations that might make development of such platform difficult but also issues connected to policy and governance issues, data reliability and trustworthiness, and creation of economic incentives for various actors along the supply chain. In the concluding section of this chapter, we discuss some of the practical implications of these challenges in regard to creating conditions that would allow for successful integrated data sharing along the supply chain for sustainable coffee.

2.5 Conditions for Developing Interoperable Data Platform in Sustainable Supply Chain

An interoperable data platform does not consist just of the software, hardware, and data contained in it. Rather it is a system enveloping components of governance, policy, business, trustworthiness, stakeholder relations, and more. The following paragraphs begin to sketch a few system components and conditions that are necessary for a good foundation of an interoperable data platform.

2.5.1 *Ensuring Information Integrity, Trustworthiness, and Security*

The functionality and usefulness of any information sharing system depends on the degree to which information contained in the system is accurate, timely, and useful and the extent to which it protects integrity of the said information. In other words, assuring data's trustworthiness is crucial when data is being shared among various sources and in various formats and also in creating aggregated information from the shared data (Dinh, Wenqiang, & Datta, 2012). The degree of trust toward the data and the platform's security presumably influences the likelihood that both consumers and supply chain partners will use the information produced by a system like I-Choose, and thus creating technical and process mechanisms to ensure information integrity is essential.

Information security in platforms such as I-Choose is particularly important because data that is relevant might often be considered confidential and/or proprietary. As noted in previous sections, our interviewees considered some of the information disclosed during a certification process as vital to their economic competitiveness. As a result, any information architecture or platform would need to incorporate proper access controls to authenticate its diverse users to ensure that proprietary data is not being accessed inappropriately. To ensure participation of some of the key actors of the supply chain, the developers must find an effective way to involve primary stakeholders, if not primary data producers, in designing the information security requirements from the start. Such participation will boost stakeholders' trust in the data and enhance confidence among data owners that their proprietary data is well protected (see Chap. 6 for a discussion on security and privacy requirements and challenges).

In addition to data security, developers also need to ensure trustworthiness of data in their system. Recent research examined trustworthiness of data through various lenses, namely, data integrity, data quality, and data lineage and provenance (Bertino & Lim, 2011; Bertino, Dai, & Kantarcioglu, 2009). Previous research on data quality suggests that, from the user perspective, data quality correlates to high relevance (Bertino et al., 2009; Tayi & Ballou, 1998; Wang & Strong, 1996). In terms of data provenance, trustworthiness of the data is defined as having access to the source and origins of the data (Bertino & Lim, 2011; Bertino et al.; Ram & Liu, 2009).

Few researchers, however, pay attention to the role that can be played by governments. Presumably, government policy promoting disclosure of information held not only by public entities but also private entities contributes to market transparency and trust production in sustainable markets. Trust is often related to institutional factors, such as legal contracts, social networks, and societal norms that make opportunistic behaviors less likely. Arguably, government efforts in establishing legal framework to ensure consistent product standards for sustainable products could contribute to the development of consumer trust. Using a survey, we investigated the determinants of consumer trust in the presumed sustainability of a product (Sayogo, Zhang, Liu, Picazo-Vela, & Luna-Reyes, 2014). Our results suggest that in the case of sustainable products, brands, certificate's reputation, support from government agencies and endorsement from nonprofit organizations significantly influence consumers' trust in the product. We argue that to be useful, additional information from the certification process should be aggregated and presented to consumers in a simple and value-added fashion.

Nonetheless, our research is only the first step to understanding the dynamic interaction of market and government policy. The issues of generating and measuring data trustworthiness from the user perspective are still very much open for thorough investigation. The challenge of creating trustworthiness in the system is discussed in more detail in Chap. 3.

2.5.2 Creation of Semantic Compatibilities Among Standards and Protocols

Although the interviewees did not explicitly emphasize the issue of semantic incompatibilities, this issue is crucial for developing platform such as I-Choose. There are diverse certification standards and protocols applied to the sustainable certified coffee supply chain (van Hoek, Vos, & Commandeur, 1999; Wang & Strong, 1996), and each uses different criteria and processes for assessment and certification. The EcoLabelIndex,[4] an information aggregator on eco-certifications, is currently tracking 444 ecolabels in 197 countries and 25 industry sectors. Consequently, to provide trustworthy recommendations, platforms such as I-Choose need to take into account these diversities by creating semantic compatibilities among standards and protocols. Technically, the development of semantic compatibilities and common language and schema among diverse knowledge is possible through ontology to facilitate automatic data extraction and reasoning about sustainable certification and trading. The primary stakeholders in sustainable certified coffee supply chain could share a common understanding of the certification structures through the ontology layers. Chapter 5 in this book includes a more detailed description of our project's efforts toward developing an ontology that would accommodate data from sustainable coffee supply chain.

2.5.3 Designing Information Policy That Balances Commercial Interests and Openness

One of the biggest challenges facing a system such as I-Choose is the lack of economic motivations for information disclosure. The interviewees indicated inclination and willingness to open their information only if doing so adds value to their organizations. Thus, it is important to design information policy that balances the desire for supply chain transparency and the need for keeping businesses competitive. A lesson from studies in economics can be used as a starting lens. These studies point out that timing and manner of disclosure supersede the decision of whether companies need to disclose. Appropriate timing and method for information disclosure can mitigate the company's fears that disclosure will disrupt their market position and thus allow them to accrue the benefits from disclosing information (Vachon & Klassen, 2007).

[4] http://www.ecolabelindex.com

2.5.4 Creation of Business Model for Integrated Data Platform

The potential for a business or revenue model of systems like I-Choose has both practical and research implications. Arguably, the existence of an integrated and open data architecture might provide fertile ground for entrepreneur development. Processing and providing users' information that fits their needs could create innovative business models based on market signals and value provisions.

Our current simulation experiments reveal that the market resists "takeoff" unless external financial support can be found (Ran et al., 2016). Additionally, "takeoff" dynamics of the system are dominated by marketing budgets and external support for infrastructure. Marketing budgets drive how fast users adopt the system, and without external sponsorship of system, the final market collapses. Further research investigating an appropriate business model would be beneficial for the development and adoption of such platforms such as I-Choose.

2.5.5 Establishing Collaborative Governance Model

A number of the challenges discussed above call for attention to policy development in respect to governance. Specifically, interoperable systems that integrate various organizations in the fragmented supply chain and multiple certifications would involve multiple stakeholders with diverse interests, degrees of power, and values. Integration of these diverse values necessitates the construction of governing mechanisms that take these complexities into account.

One possible governance intervention is the creation of data commons governed by collaborative governance body (Flynn, Huo, & Zhao, 2010). This is a roundtable type of governance model involving stakeholders in the supply chain. The function of such governance mechanism is to generate policies to support the implementation of platform such as I-Choose. This governance body supervises the creation of semantic compatibilities among standards and protocols, the creation of information security policies, and the design of information policy that balances the need for commercial privacy and desire for information openness. Jarman et al. (2011) envisioned balancing three elements of (1) "hard" regulation, (2) partnership building, and (3) wider participation from consumer review as one plausible governing mechanism for I-Choose. Chapter 8 of this book further discusses governance issues and further paths for development.

The involvement of various supply chain stakeholders including government regulators, industry associations, consumers, consumer advocates, producers, and others is a must. Involvement of these stakeholders in the governance body minimizes resistance caused by variety of conflicting interests as well as making sure that all stakeholders' interests are represented. This way the issue of negative impacts to sustainability and competitiveness can be mitigated, especially for stakeholders with less power and influence.

2.6 Concluding Remarks

Information asymmetry in the relationship between end consumers and firms in supply chain is argued to be one of the key barriers that confine the proliferation of sustainable consumption (Seyfang, 2005). A challenge to remedy such problems lies in making vast amounts of disparate data and information regarding sustainability practices shareable across the supply chain and usable by end consumers. One key missing element is a platform that combines interoperable data standards and architecture with policy and governance mechanisms. This chapter presents main challenges and key conditions to be considered in developing data standards to support interoperable platform for sustainability as perceived by key stakeholders in sustainable certified coffee industry.

References

Akerlof, G. A. (1970). The market for "lemons": Quality uncertainty and the market mechanism. *The Quarterly Journal of Economics, 84*(3), 488–500.

Angell, L. C., & Klassen, R. D. (1999). Integrating environmental issues into the mainstream: An agenda for research in operations management. *Journal of Operations Management, 17*(5), 575–598. http://doi.org/10.1016/S0272-6963(99)00006-6.

Beamon, B. M. (1999). Designing the green supply chain. *Logistics Information Management, 12*(4), 332–342.

Bertino, E., & Lim, H. S. (2010). Assuring data trustworthiness-concepts and research challenges. In Secure Data Management (pp. 1–12). Springer Berlin Heidelberg.

Bertino, E., Dai, C., & Kantarcioglu, M. (2009, January). The challenge of assuring data trustworthiness. In Database Systems for Advanced Applications (pp. 22–33). Springer Berlin Heidelberg.

Clapp, J. (1998). The privatization of global environmental governance: ISO 14000 and the developing world. *Global Governance, 4*(3), 295–316.

Clemons, E. K., & Row, M. C. (1993). Limits to interfirm coordination through information technology: Results of a field study in consumer packaged goods distribution. *Journal of Management Information Systems, 10*(1), 73–95.

Collins, C. M., Steg, L., & Koning, M. A. S. (2007). Customers' values, beliefs on sustainable corporate performance, and buying behavior. *Psychology and Marketing, 24*(6), 555–577.

Darby, M. R., & Karni, E. (1973). Free competition and the optimal amount of fraud. *Journal of Law and Economics, 16*(1), 67–88.

Dinh, T. T. A., Wenqiang, W., & Datta, A. (2012). City on the sky: Extending XACML for flexible, secure data sharing on the cloud. *Journal of Grid Computing, 10*(1), 151–172. http://doi.org/10.1007/s10723-012-9212-9.

Eisenhardt, K. M. (1989). Agency theory: An assessment and review. *The Academy of Management Review, 14*(1), 57–74.

Fama, E. F., & Jensen, M. C. (1983). Separation of ownership and control. *Journal of Law and Economics, 26*(2), 301–325.

Flynn, B. B., Huo, B., & Zhao, X. (2010). The impact of supply chain integration on performance: A contingency and configuration approach. *Journal of Operations Management, 28*(1), 58–71. http://doi.org/10.1016/j.jom.2009.06.001.

Fuchs, D. A., & Lorek, S. (2005). Sustainable consumption governance: A history of promises and failures. *Journal of Consumer Policy, 28*(3), 261–288. http://doi.org/10.1007/s10603-005-8490-z.

Goleman, D. (2010). Ecological intelligence: The hidden impacts of what we buy. Broadway Business.
Han, K., Chang, Y. B., & Hahn, J. (2011). Information technology spillover and productivity: The role of information technology intensity and competition. *Journal of Management Information Systems, 28*(1), 115–146. http://doi.org/10.2753/MIS0742-1222280105.
Hardin, G. (1968). The tragedy of the commons. *Science, 162*(3859), 1243–1248.
Howard, A. (2012, April 1). *What is smart disclosure?* Retrieved from http://radar.oreilly.com/2012/04/what-is-smart-disclosure.htm
Hsu, C.-C., Tan, K. C., Zailani, S. H. M., & Jayaraman, V. (2013). Supply chain drivers that foster the development of green initiatives in an emerging economy. *International Journal of Operations & Production Management, 33*(6), 656–688. http://doi.org/10.1108/IJOPM-10-2011-0401.
Jahn, G., Schramm, M., & Spiller, A. (2005). The reliability of certification: Quality labels as a consumer policy tool. *Journal of Consumer Policy, 28*(1), 53–73. http://doi.org/10.1007/s10603-004-7298-6.
Jarman, H., Luna-Reyes, L. F., Zhang, J., Whitmore, A., Picazo-Vela, S., Andersen, D. L., et al. (2011). *I-choose: Consumer choice, digital government, and sustainability in North America.* Presented at the APPAM Fall Research Conference, Washington D.C.
Johnston, H. R., & Vitale, M. A. (1988). Creating competitive advantage with interorganizational information systems. *MIS Quarterly, 12*(2), 153–165.
Linton, J. D., Klassen, R., & Jayaraman, V. (2007). Sustainable supply chains: An introduction. *Journal of Operations Management, 25*(6), 1075–1082.
Locke, R., & Romis, M. (2007). Improving work conditions in a global supply chain. *MIT Sloan Management Review, 48*(2), 54–61.
Locke, R., Kochan, T., Romis, M., & Qin, F. (2007). Beyond corporate codes of conduct: Work organization and labour standards at Nike's suppliers. *International Labour Review, 146*(1–2), 21–40.
Malhotra, A., Gosain, S., & Sawy, O. A. E. (2005). Absorptive capacity configurations in supply chains: Gearing for partner-enabled market knowledge creation. *MIS Quarterly, 145–187.*
Matos, S., & Hall, J. (2007). Integrating sustainable development in the supply chain: The case of life cycle assessment in oil and gas and agricultural biotechnology. *Journal of Operations Management, 25*(6), 1083–1102. http://doi.org/10.1016/j.jom.2007.01.013.
Mishra, D. P., Heide, J. B., & Cort, S. G. (1998). Information asymmetry and levels of agency relationships. *Journal of Marketing Research, 35*(3), 277. http://doi.org/10.2307/3152028.
Mont, O., & Plepys, A. (2008). Sustainable consumption progress: Should we be proud or alarmed? *Journal of Cleaner Production, 16*(4), 531–537. http://doi.org/10.1016/j.jclepro.2007.01.009.
Nelson, P. (1970). Information and consumer behavior. *Journal of Political Economy, 78*(2), 311–329.
Opara, L. U. (2003). Traceability in agriculture and food supply chain: A review of basic concepts, technological implications, and future prospects. *Journal of Food, Agriculture and Environment, 1*, 101–106.
Ram, S., & Liu, J. (2009). A New Perspective on Semantics of Data Provenance. SWPM, 526.
Ran, W., Jarman, H., Luna-Reyes, L. F., Zhang, J., Andersen, D. L., Tayi, G. K., et al. (2016). Supply-chain transparency and governance systems: Market penetration of the I-choose system. In J. Zhang, L. F. Luna-Reyes, T. A. Pardo, & D. S. Sayogo (Eds.), *Information, models, and sustainability: Policy informatics in the age of big data and open government*. Pravi: Springler.
Sarkis, J., Zhu, Q., & Lai, K. (2011). An organizational theoretic review of green supply chain management literature. *International Journal of Production Economics, 130*(1), 1–15. http://doi.org/10.1016/j.ijpe.2010.11.010.
Sayogo, D. S. (2013). *Modeling incentives to disclose: Smart disclosure policy, private sector transparency and demanded disclosure* (Dissertation). University at Albany State University of New York, Albany.
Sayogo, D. S., Jarman, H., Whitmore, A., Tayi, G. K., Zhang, J., Hrdinova, J., ... & Andersen, D. L. (2012, June). *A stakeholder analysis of interoperable data architecture*: The case of I-Choose. In Proceedings of the 13th Annual International Conference on Digital Government Research (pp. 145–154). ACM.

Sayogo, D. S., Zhang, J., Liu, H., Picazo-Vela, S., & Luna-Reyes, L. (2014, June). *Examining trust as key drivers in smart disclosure for sustainable consumption*: the case of I-choose. In Proceedings of the 15th Annual International Conference on Digital Government Research (pp. 137–146). ACM.

Sayogo, D. S., Zhang, J., Pardo, T. A., Tayi, G. K., Hrdinova, J., Andersen, D. F., et al. (2014). Going beyond open data: Challenges and motivations for smart disclosure in ethical consumption. *Journal of Theoretical and Applied Electronic Commerce Research, 9*(2), 3–4. http://doi.org/10.4067/S0718-18762014000200002.

Sayogo, D. S., Zhang, J., Luna-Reyes, L., Jarman, H., Tayi, G., Andersen, D. L., et al. (2015). Challenges and requirements for developing data architecture supporting integration of sustainable supply chains. *Information Technology and Management, 16*(1), 5–18. http://doi.org/10.1007/s10799-014-0203-3.

Senge, P. M., Smith, B., Kruschwitz, N., Laur, J., & Schley, S. (2008). The necessary revolution: How individuals and organizations are working together to create a sustainable world. Crown Business.

Seyfang, G. (2005). Shopping for sustainability: Can sustainable consumption promote ecological citizenship? *Environmental Politics, 14*(2), 290–306. http://doi.org/10.1080/09644010500055209.

Steinfield, C., Markus, M. L., & Wigand, R. T. (2011). Through a glass clearly: Standards, architecture, and process transparency in global supply chains. *Journal of Management Information Systems, 28*(2), 75–108. http://doi.org/10.2753/MIS0742-1222280204.

Tayi, G. K., & Ballou, D. P. (1998). Examining data quality. *Communications of the ACM, 41*(2), 54–57. http://doi.org/10.1145/269012.269021.

Thaler, R. H., & Sunstein, C. R. (2008). *Nudge: improving decisions about health, wealth, and happiness*. New Haven: Yale University Press.

Vachon, S., & Klassen, R. (2007). Supply chain management and environmental technologies: the role of integration. *International Journal of Production Research, 45*(2), 401–423.

van Hoek, R. I., Vos, B., & Commandeur, H. R. (1999). Restructuring European supply chains by implementing postponement strategies. *Long Range Planning, 32*(5), 505–518. http://doi.org/10.1016/S0024-6301(99)00071-0.

Wang, E., Tai, J., & Wei, H.-L. (2006). A virtual integration theory of improved supply-chain performance. *Journal of Management Information Systems, 23*(2), 41–64. http://doi.org/10.2753/MIS0742-1222230203.

Wang, R. Y., & Strong, D. M. (1996). Beyond accuracy: What data quality means to data consumers. *Journal of Management Information Systems, 12*(4), 5–33.

Wilson, T. P., & Clarke, W. R. (1998). Food safety and traceability in the agricultural supply chain: Using the Internet to deliver traceability. *Supply Chain Management: An International Journal, 3*(3), 127–133. http://doi.org/10.1108/13598549810230831.

Wolf, J. (2011). Sustainable supply chain management integration: A qualitative analysis of the German manufacturing industry. *Journal of Business Ethics, 102*(2), 221–235. http://doi.org/10.1007/s10551-011-0806-0.

Zhu, Q., & Sarkis, J. (2004). Relationships between operational practices and performance among early adopters of green supply chain management practices in Chinese manufacturing enterprises. *Journal of Operations Management, 22*(3), 265–289. http://doi.org/10.1016/j.jom.2004.01.005.

Chapter 3
Collaboration and Trust Building Among Public and Private Actors

Luis F. Luna-Reyes, Deborah Lines Andersen, David F. Andersen, and Holly Jarman

Abstract Using data from the I-Choose project, a study of coffee produced in Mexico and distributed and sold in Canada and the United States, this chapter analyzes three distinct traceability systems in relation to the ways in which each attempts to build and sustain trust. In each case, supply chain actors are working together to capture information about how and where their products are produced, aiming to provide this information to consumers. The ultimate goal in each system is the same: to demonstrate the quality of their product and earn a price premium. We find that institutional, calculative, and relational trust are used in different ways in each of the three systems, with distinct variations over time. Extrapolating from these cases, we find that providing consumers with sustainable supply chain information evolves dynamically over time with calculative trust less permanent and relational trust more permanent. Institutional trust appears to be the best way to communicate with consumers in international marketplaces.

Keywords Trust • Provenance • User confidence • Information assurance • Social computing • Customer-oriented systems

L.F. Luna-Reyes (✉)
Department of Informatics, University at Albany, Albany, NY, USA
e-mail: lluna-reyes@albany.edu

D.L. Andersen
Department of Information Studies, University at Albany, Albany, NY, USA
e-mail: dla@albany.edu

D.F. Andersen
Rockefeller College of Public Affairs and Policy, University at Albany, Albany, NY, USA
e-mail: david.andersen@albany.edu

H. Jarman
Department of Health Management and Policy, School of Public Health, University of Michigan, Ann Arbor, MI, USA
e-mail: hjarman@umich.edu

H. Jarman, L.F. Luna-Reyes (eds.), *Private Data and Public Value*, Public Administration and Information Technology 26,
DOI 10.1007/978-3-319-27823-0_3

3.1 Introduction

In a period in which more and more product information is available electronically, it can still be extremely difficult to understand the provenance of a product. Origins and history, from farmer to final customer for coffee, can be impossible to follow to the final point of purchase. Despite online reviews and internationally recognized labeling systems, customers often know relatively little about products that they buy, especially in highly dispersed, international markets and supply chains. Individually, farmers, manufacturers, distributors, certifiers, and retailers have some understanding of their part in the supply chain. However, it is highly likely that none of them would be able to map the supply chain in its entirety. If the ultimate goal of a system is to demonstrate the quality of a product and earn a price premium, then providing customers trusted information on provenance is a critical factor in making a traceability system a viable marketing tool.

As the previous chapter explains, understanding every aspect of a product supply chain is necessary if we are to support sustainable commerce and perhaps design more sustainable supply chains. More and more commerce is taking place in virtual settings where product attributes cannot be directly observed by the consumer (Ba & Pavlou, 2002). Although consumers might be drawn to farmers' markets in which the supply chain is extremely short—a producer and a customer—much of what consumers purchase has a much more complicated history.

Unequal information distribution between consumers and producers appears to be a fundamental issue across supply chains. Consumers often know much less about products than do producers, although it is very possible that information lost across the supply chain means that farmers, manufacturers, distributors, certifiers, and retailers all know only a portion of a where a product has been, how it has been produced, and by whom under what conditions. This unequal distribution of information, described by Akerlof (1970) in his discussion of used car markets, creates a lack of trust on the part of consumers. How is it possible for a consumer to believe what a retail package says? How might stakeholders along the supply chain come to trust the information that they are given and then supply trusted information as a product moves toward the consumer? How can institutional mechanisms such as guarantees, certifications, brands, and supply chains reduce the uncertainty associated with information asymmetries in the market (Backhouse, Hsu, Tseng, & Baptista, 2005; Pan, 2011; Pavlou, Liang, & Xue, 2007; Sriram, 2005)?

Assembling and assessing supply chain data with a view to supporting sustainability and influencing customer purchases requires something more than coercion alone. In order for producers, roasters, and retailers to invest in the collection and open sharing of data—often sensitive commercial information—a significant degree of trust must be established. Consumers, too, must find credible the supply chain information provided to them if their purchasing preferences are to be influenced and changed (Arora, 2006). In traditional supply chains that emphasize price above all other values, trust among organizations is often established through rational calculations based on economic data and supplemented by institutional mechanisms such as contracts. Full Information Product Pricing (FIPP) systems, however, emphasize not only the value of price and cost efficiency but also attempt to communicate information to consumers regarding other aspects of the supply chain such as sustainable, environmentally friendly processes, and fair labor practices. This more complex goal arguably requires a test of trust-building mechanisms with a slightly different emphasis.

This chapter asks, given these conditions, how is trust built and sustained among sustainable supply chain actors with different levels of information? Using data from the I-Choose project, which focuses on goods produced in Mexico and distributed and sold in the United States and Canada, we review three distinct traceability systems: Tosepan Titataniske (Together We Win), a cooperative that exports organic and fair-trade coffee to the United States, Japan, and Europe; a coffee roaster in Chiapas, Mexico, named Bats'il; and the Nescafé plan, led by Nestlé. Each of these systems approaches the problem of information asymmetry along the supply chain in a different way. All three, however, share the same objective of capturing data about the provenance of the products—information about where, how, and under what conditions they are created, distributed, and sold—in order to pass this information on to the consumer. In each case, producers hope to collect and communicate valuable data that will generate consumer trust in the product and support a price premium.

After this introduction, the chapter continues with a discussion of three overarching types of trust creation mechanisms. Following an explanation of methodology, data, and case selection criteria, the chapter finishes by comparing the three cases and the ways in which trust is generated

3.2 Creating and Maintaining Trust

There appears to be no generally accepted definition of trust (Rousseau, Sitkin, Burt, & Camerer, 1998; Sheppard & Sherman, 1998). Nonetheless, there are some common themes and general understandings of its nature, including vulnerability, risk, and the role of positive expectations or optimistic belief (Rousseau et al., 1998; White-Cooper, Dawkins, & Anderson, 2009). Trust is usually seen as a two-party relationship, in which one party A (an individual or an organization) accepts the inherent risk of a relationship with another party B (Sheppard & Sherman, 1998). In fact, some authors point out that different kinds of relationships between the same participants can have different levels of risk. In this way, A can trust B for taking care of his or her dog, but might not trust B for doing business (Viitaharju & Lähdesmäki, 2012).

We find that each of these traceability systems attempts trust production in three ways that correspond to established theories of trust building. According to Rousseau et al. (1998), these three mechanisms are interrelated. For example, institutional mechanisms of trust creation (1) relate to calculative mechanisms by reducing the perception of risk associated with a particular transaction or relationship.

Calculative trust (2) plays a more important role at the beginning of a relationship, and after repeated positive interactions, greater elements of relational trust (3) arise and play a more important role in the interaction between individuals and organizations.

3.2.1 *Institutional Trust*

First, trust is built through institutions such as contracts, formal agreements, certification procedures, laws, and regulations (Burkert, Ivens, & Shan, 2012). These institutions come together to create a trusted environment that can facilitate the development of other types of trusted systems. Institutional trust refers to the existence of an institutional framework that regulates the relationship between the main actors in the collaboration. These institutional frameworks include the basic institutional mechanisms discussed by Akerlof (1970) such as contracts, guarantees, and certifications (Albersmeier, Schulze, Jahn, & Spiller, 2009; Jahn, Schramm, & Spiller, 2005), as well as the laws, regulations, and other institutions oriented to enforce the contracts. Research has found that institutional trust is particularly relevant for systems such as I-Choose and also that some features of information systems and information technologies contribute to building institutional trust (Gefen, Pavlou, Benbasat, McKnight, & Stewart, 2006; Thornton, Esper, & Morris, 2013). Some of these features include peer feedback, online testimonials, affiliation links, guarantees, or system quality. In each case there are social systems behind these system features that make them legitimate.

3.2.2 *Calculative Trust*

Next, by identifying information about available options and weighing costs and benefits, supply chain actors can manage the risks inherent in participating in any one system. Calculative trust refers to an estimation of the risks and payoffs intertwined in the interaction. Stakeholders may conceptualize a ratio of payoffs and risks for each interaction between individuals and organizations. This ratio is conditional and only arises when reliable information attests to the beneficial intention and competence of another player in the relationship (Rousseau et al., 1998).

3.2.3 *Relational Trust*

Third and finally, relational trust is established through emotional bonds, shared values or objectives between the actors, or recognition of the trustworthiness of other participants in a repeated relationship. Relational trust is associated with emotional bonds, shared values or objectives between the actors, or recognition of the trustworthiness of other participants in a repeated relationship. Compared to calculative trust, relational trust is more resilient to environmental changes and may cover a broad array of interactions (Román, 2010; Saini, 2010; Saparito, Chen, & Sapienza, 2004; Smith & Barrientos, 2005; Zornoza, Orengo, & Penarroja, 2009).

3.3 Methods, Data, and Case Selection

As introduced in Chap. 1, this book reports on the I-Choose project. The main purpose of the project was to better understand how information disclosure might influence firm and consumer behavior and potentially create a more sustainable world. Following a case study approach, this chapter now moves to report on a project component that consisted of further understanding the ways in which trust is built and sustained among supply chain actors (Eisenhardt, 1989; Stake, 1995; Yin, 1994). Given the focus on process and the emphasis on "how" trust is built, the case study is an appropriate methodological approach (Yin, 1994).

Case selection obeyed the criteria of theoretical sampling (Eisenhardt, 1989; Eisenhardt & Graebner, 2007). We used the conceptual framework introduced in the previous section to analyze a series of collaborative traceability efforts in Mexico. For our purposes, "collaborative traceability efforts" have three key characteristics. They are (i) collaborations among coffee supply chain actors, (ii) with the purpose of gathering information about the provenance and quality of their products and communicating this information to consumers, (iii) in ways that support a price premium for those coffee products. For each of the selected collaborative traceability efforts, we were interested in the mechanisms by which supply chain actors attempted to build trust—both among the collaborators themselves and between the collaborators and consumers.

Our initial survey of coffee suppliers in Mexico revealed wide diversity among the population of collaborative traceability efforts. After identifying initial contacts for interview, we used a snowball sampling method to identify further interviewees. We conducted 36 interviews in two different time periods, interviewing participants in the coffee supply chain as well as certifiers. As a result, we identified several supply chain systems introduced in Chap. 2. While some systems rely heavily upon institutions to generate trust, others rely almost exclusively on building close relationships. Among this diversity, we selected the three cases that were more theoretically rich in terms of trust relationships. Sixteen out of the 36 interviews were related to these three final cases.

The interview protocol was developed and revised by an interdisciplinary team of researchers and included the following topics: the mission and objectives of the organization represented by the interviewee, an exploration of information that is currently shared by their organization as well as information that should be shared, the level of trust in the information provided, barriers and incentives to sharing data, and collaboration, governance, and certification processes. The interview protocol was translated into Spanish by native speakers and then retranslated back into English in order to test for consistency in the terms used. Interviews were conducted in Spanish by native speakers and formally transcribed.[1]

Data analysis included the identification of key themes in the data as well as a process of comparing and contrasting both cases and current theory in an iterative process (Eisenhardt, 1989). Following common practices in qualitative research, data analysis started during the data collection process, allowing us to identify the gaps in the available data so that they could be addressed in subsequent data collection activities. The validity of constructs used in our analysis comes from their currently accepted use in the literature.

3.4 Collaborative Traceability Efforts Within Mexican Coffee Supply Chains

In this section of the chapter, we include a description of each of the three cases of collaborative traceability efforts, analyzing them in terms of the trust-building mechanisms employed by each. The cases are (1) a coffee cooperative in the mountains of Puebla, Mexico, named Tosepan Titataniske; (2) a coffee roaster in Chiapas, Mexico, named Bats'il; and (3) the Nescafé plan, led by Nestlé.

3.4.1 Tosepan Titataniske (Together We Win)

As a response to the coffee market crisis of the 1990s, coffee producers in Mexico adopted the concept of fair trade. Fair trade involves a series of quality, organic, and social standards devised to differentiate coffee produced under these norms, increasing sales price and reducing risks of price fluctuations. Tosepan Titataniske is a cooperative in the northern mountains of the state of Puebla in Mexico, which produces and exports organic and fair-trade coffee to the United States, Japan, and Europe. The Tosepan cooperative groups about 1400 small producers from about 70 communities in the mountains. Tosepan is organized as a network of local cooperatives that collaborate to sell coffee through a central warehouse at Cuetzalan, the main city in the area.

Tosepan is certified as an organic/fair-trade coffee producer by Fair Trade Mexico, Certimex, Ocia International, and Fair Trade Labeling Organization (FLO). The certifying process involves inspection of local small producers by visiting their lands and establishing production quotas for each of them. The total amount of organic or fair-trade coffee that Tosepan can sell/export is the sum of each small producer's quota.

[1] This research was supported by CONACYT grant no. 133670.

Tosepan has a manual traceability system to control individual quotas. Although they use the Internet (e-mail) and some basic productivity applications (spreadsheets and word processing), information technologies have the potential to facilitate certification and traceability of coffee in the network of producers. However, the Tosepan interviewee showed caution about the use of a traceability system like the one described in the interview policy scenario. His reaction was "we like systems to have traceability and transparency, but when a system is big, it can be heavy as a rock and it needs to be carried.... [I]f the system is too rigid, it can leave out many possibilities to producers."

Moreover, and according to Tosepan's interviewee, fair-trade exports could benefit from having clearer government standards and regulations, which are much more developed for organic products. However, Tosepan's relationships with the government have been limited and difficult. As the informant expressed, "Today government is interested in organic and fair trade because of the market, and not the philosophy. They have realized that conventional producers are out of business. They did not believe all people who approached them before. Our relations with government have been complicated.... Although there are government officials that show a lot of interest in their work, there is a huge bureaucracy that makes it hard for us to see government as our partner."

Institutional Trust Tosepan producers rely heavily on institutional mechanisms to introduce their products in the international market. Organic and fair-trade seals provided by Certimex, Ocia International, and FLO are important instruments to inform coffee producers and consumers in the international markets about the environmental and social practices at Tosepan. Their current traceability system not only is a requirement from all certification authorities but also a tool to promote transparency and trust among members of the cooperative reducing opportunistic behaviors such as paying friends fair-trade or organic prices when they do not meet the requirements.

Calculative Trust The robustness and strength of Tosepan's institutional mechanisms help to build calculative trust. This trust plays a role when starting a relationship with a new client (importers or exporters) who relies on Tosepan's capabilities to meet the time requirements of a transaction. However, even in those initial processes, relational trust has proven more important as we will discuss in the following section.

Relational Trust Relational trust plays an important role in Tosepan's success. In fact, through its relationships with other Mexican cooperatives, Tosepan participates in a number of organizations promoting fair trade in Mexico and abroad. Some of these organizations include Comercio Justo Mexico, Agromercados, and Certimex. This network of organizations has been key in developing marketing and commercial capabilities among cooperatives.

More specifically, the network (relational trust) has been important for building export capabilities in Tosepan through its relationships with other fair-trade cooperatives in Mexico. According to the interviewee, these relationships were key to their ability to start exporting, and currently they are helping other cooperatives to enter the market by selling their products.

Moreover, it has also played an important role in the creation of contacts with buyers. As the interviewee commented, "Exporting is difficult because of the language and the need to learn many rules and terms. Selling to a national broker has its intricacies, but it is easier. Exporting requires more responsibility during the negotiation, logistics and follow-up. We have learned slowly, and we have been able to learn because of conscientious buyers who know that it is difficult and have helped us to do it."

3.4.2 Bats'il Maya

Azúcar Morena (raw sugar) and Capeltic are two specialty coffee shops in central Mexico. Both coffee shops constitute the front end of a network of organizations promoting organic and fair-trade coffee in Mexico and abroad. At the heart of this network is Bats'il Maya, a gourmet coffee roasting organization established in 1993 in Chilon, Chiapas, Mexico. Similar to Tosepan Titataniske, Bats'il Maya started as a way to deal with the coffee crisis of the 1990s. However, Bats'il Maya is not a cooperative, but a micro-roaster founded by a Jesuit Mission in Chiapas. Although Bats'il Maya exports coffee to Japan, it also collaborates with a very well-known coffee taster, local cooperatives from the area, and specialty coffee shops such as Capeltic and Azúcar Morena to integrate the supply chain "from the plant to the cup." The coffee mix that consumers can taste in each coffee shop is unique.

Although Bats'il is—according to a local certifier—an organic certified organization, coffee shops like Azúcar Morena and Capeltic do not use the organic label. Both coffee shops, nevertheless, claim to sell organic coffee from Chiapas. The certification of the coffee was not even mentioned during the interview and is not included in any of the organizations' websites. The manager of Azucar Morena mentioned several times that the coffee taster working in Bats'il was a well-recognized taster offering technical assistance to local producers to continuously improve coffee production processes. Every claim was also supported by extensive explanations of the process and the local cooperatives involved in the supply chain. Capeltic is located at Jesuit colleges in Mexico, which may generate trust among consumers.

All organizations involved in the network rely heavily on personal relations for trade. Moreover, all share a special interest to promote Mexican products and other ethical and aesthetic values. For instance, Azucar Morena combines gourmet coffee with a space that can be used as an art gallery by local artists. The owner, recognizing the special value of coffee for the cooperatives in Chiapas, attempts to build a bridge between Indian communities and other Mexicans.

Institutional Trust Institutional trust is not an important source of trust in this specific coffee supply chain. For consumers at Capeltic, maybe the only source of institutional trust comes from the Jesuit endorsement of both Bats'il and the coffee shops. In the case of Azucar Morena, the main source of institutional trust is perhaps the personal accreditations of the coffee taster working at Bats'il.

Calculative Trust From our point of view, there is no evidence of calculation as a source of trust among supply chain participants. Maybe the only source of calculative trust is the premium paid to producers in exchange for high-quality coffee or the expectation of increased profits by selling gourmet coffee.

Relational Trust Relationships may be the most important sources of trust among this network of organizations. As the owner of Azucar Morena commented, "I started this coffee shop because I am a friend of the coffee taster at Bats'il, and saw him at another coffee shop that was already brewing its own coffee mix." Moreover, Bats'il has a close relationship with local coffee cooperatives, buying coffee from them at a fair price and also providing technical assistance to continuously improve the quality of the product. Jesuits relationships have been also a key element in the creation of some coffee shops such as Capeltic.

3.4.3 The Nescafé Plan

Nestlé is a global company with headquarters in Switzerland. Company history can be traced back to 1866 with the opening of their first condensed milk factory in Cham, Switzerland. Since then, the organization has grown and diversified its operations to become the global leader in the food industry. Particularly referring to coffee, Nestlé created the Nescafé brand in the 1930s in Brazil and currently is one of the leading global brands of the company. In Mexico, Nestlé is the leader in coffee roasting and production, and it owns the biggest coffee factory in the world, located in Toluca, Mexico. In 2010, Nestlé launched the Nescafé plan as a global program to promote sustainability in the coffee industry as well as ethical consumption. The program includes partnerships with coffee producers and other supply chain participants, as well as a traceability program to better communicate quality to consumers. With a global investment of 500 million Swiss francs, Mexico is a pilot country for the program.

Nestlé is a corporate partner of the Common Code for the Coffee Community or the 4C program, consisting of a third-party verification of best practices in terms of quality, social, and environmental principles. The 4C principles and practices are the result of the collaboration of some national governments as well as standard-setting organizations such as UTZ and the Rainforest Alliance. However, 4C does not involve a label, but a commitment of compliance with their basic principles and practices. This compliance is verified by third-party auditors. As of fall 2013, Nestlé bought coffee from 12,000 4C-compliant producers. The goal is to reach out to 70,000 producers through the program by 2015.

One important component of the Nescafé plan is the construction of a traceability system for coffee that will allow Nestlé to communicate to consumers about coffee origin and quality. In México, Nestlé has collaborated directly with medium and large producers, and with large intermediaries, to gather information from small producers. Although building a database has been a quite successful project—there are already 80,000 producers in the system—using the information has proved a challenge. As one of the interviewees commented, "One single coffee lot for the production process includes coffee from 300 to 400 producers, thus making it hard to trace back an individual jar of Nescafé to one producer." Nevertheless, the system is used not only for traceability purposes but also to work on quality control and to measure the impact of the Nescafé plan's benefits to producers.

From the point of view of the interviewees, sustainability should be a shared commitment along the supply chain, and promoting more sustainable systems results from partnerships with producers, governments, and NGOs. In fact, as one of the interviewees explained, "The Nescafé plan involves a group of partnerships and alliances, including the producer themselves, intermediaries, Agromor,[2] the 4C Association, and the Tec of Monterrey."[3]

Institutional Trust Nestlé main source of institutional trust is the brand itself and its long history as a world leader in the food industry. The brand, however, triggers different and contradictory perceptions among other coffee supply chain participants. In general, large roasters as well as intermediaries recognize Nestlé because of its quality control systems. For example, one large intermediary commented, "Nestlé is a very attractive client to us. We have several years working with them, and it is a company always interested in quality and innovation." However, Nestlé perception among small producer organizations is not always as positive. A second source of institution trust may, indeed, include becoming a part of the 4C Association, improving brand perception among some producer and consumer groups.

Calculative Trust Calculative trust is an important component in the Nestlé supply chain. On the one hand, being the largest coffee roaster in the country makes Nestlé a very attractive client to producers and other intermediaries. Additionally, Nestlé pays a premium price to 4C producers. As one intermediary commented, "Once you are a 4C producer, you get a premium for your coffee. A group from Tomatlán, which is already 4C, got an additional 80 cents per kg of green coffee." On the other hand, according to some interviewees, Nestlé chooses its partners on the basis of the long-term expected benefit from the relationship. That is to say, they will invest in training and improved plants mainly in those areas where they know they will get the most profit, paying less attention to areas that they calculate will

[2] Agromor is an organization that works closely with the Mexican National Institute for Agricultural Research (INIFAP) in the development of improved coffee plants. Nestlé distributes the plants to coffee producers as a component of the Nescafé plan.

[3] The Tec of Monterrey is a higher education institution that collaborates with Nestlé in training programs for producers.

not yield a high-quality product. These lower quality markets, in this sense, are out of reach of the Nescafé plan.

Relational Trust Nestlé is an organization interested in the creation of long-term relationships as part of the Nescafé plan. As we mentioned above, Nestlé invests heavily in improving the quality and production capabilities of its partners. It shares costs with intermediaries to make the program free to producers and to reach directly to medium and large producers. The relationship involves training, on-site technical assistance, and coffee plant improvement.

3.5 Comparison and Discussion of Three Trust Networks

Looking across the three cases, it is apparent that trust-building mechanisms play an important role in each, albeit in different ways. None of the collaborations rely on a single source of trust. Each network mobilized various combinations of institutional frameworks, calculating evidence as well as goodwill relationships and identity alignment to promote trust. Each of these plays overlapping as well as unique roles. Table 3.1 provides a narrative summary of the three cases and how each case relates to the three types of trust found in the literature.

An interesting aspect of the stories contained in Table 3.1 is how various types of trust evolve over time in each of the three cases. Figure 3.1 retells each of the case stories as suggestive "graphs over time." In each graph, the horizontal axis represents time, and the vertical axis is some measure of the relative prominence of each type of trust in the case narrative.

Panel A depicts an over-time story about the evolution of trust in the Tosepan Titataniske cooperative. In this case two forms of institutional trust come into play. Within the international coffee market, the fair-trade and organic labels are well known and highly trusted. The coffee market's trust in these two certifying institutions provides a basic framework for the creation and sustenance of the cooperative. On the other hand, farmer trust in the institution of the cooperative itself is not as important factor over time. Rather, high levels of institutional trust bind members of the cooperative together. Panel A illustrates that in the beginning, calculative trust played an important role as local farmers calculated their decisions to join the cooperative. However, over time the need to calculate costs and benefits of membership in the cooperative declined as overall trust in the institution of the cooperative itself grew.

Panel B depicts a quite similar dynamic story for the Bats'il Maya case. Relational trust dominates the story throughout. Players in the local market do calculate their self-interests in participating in these commercial relationships, but these calculations decline in importance over time. Institutional trust is the least important factor in this case.

The Nestlé case as depicted in panel C is somewhat more complicated. As with the Tosepan Titataniske cooperative, the international coffee market's willingness to

Table 3.1 Trust characteristics of collaborative traceability efforts in Mexico

Collaboration	Institutional	Calculative	Relational
Tosepan Titataniske: the cooperative allows over time less emphasis on long-term/ short-term calculation and more emphasis on institutional and relational trust	Provided by cooperative itself. Provides assurance that there is a premium. Reduces risk. A predictable structure to govern distribution and payment. Manages cash flow. Allows members to form more trusted relationships over time. The most important form of communication to consumers: through certification and labeling	Calculation that if sold through cooperative, there will be more benefit. Guaranteeing returns, especially when market poor. Trade-off between long-term and short-term benefit. Longer-term options require more trust to be built. Perhaps more important at first before relational trust built up	Developed over time, better trusted relationships develop with other co-op members. Reduces the importance of calculative trust over time. Members can eventually take on leadership roles and participate in decision-making
Bats'il: specialty coffee roaster that relies not on certification but on personal reputations to promote its coffee	The credentials of the taster are important in generating trust at first. Certifying authorities generate accreditation for the taster. Certification is not so important for communicating with consumers: shop does not advertised based on labels	Expectation of having coffee sold at a higher price	The relationship already exists before the coffee shop starts. The owner knows and trusts taster and trusts him/her to pick the best coffee. Taster builds relationships with producers and helps them to improve their methods and make their coffee better over time. Most important in getting new customers and building trust with consumers
Nestlé: trying to build trust through technical assistance and money. Want to produce high-quality coffee that it can sell for a premium	Brand reputation is a source of institutional trust. A good reputation for some higher-quality products. Small farmers do not trust Nestlé because they do not like the poorer-quality products	Nestlé calculates when to build relationships with intermediaries and when to partner with medium-sized farmers. Nestlé pays much better than the average. Intermediaries calculate when to sell to Nestlé: they like to sell to Nestlé because it pays more	Builds relationships with medium-sized farmers directly. Providing technical assistant to some medium-sized farmers (taking them to the second best university) to build their management, finance, and agricultural skills. They say they feel like partners in the business
	Nestle is trying to communicate to the public using institutional trust: they joined 4C (see Chap. 4)		Nestlé also builds strong relationships with intermediaries and trusts them to gather information

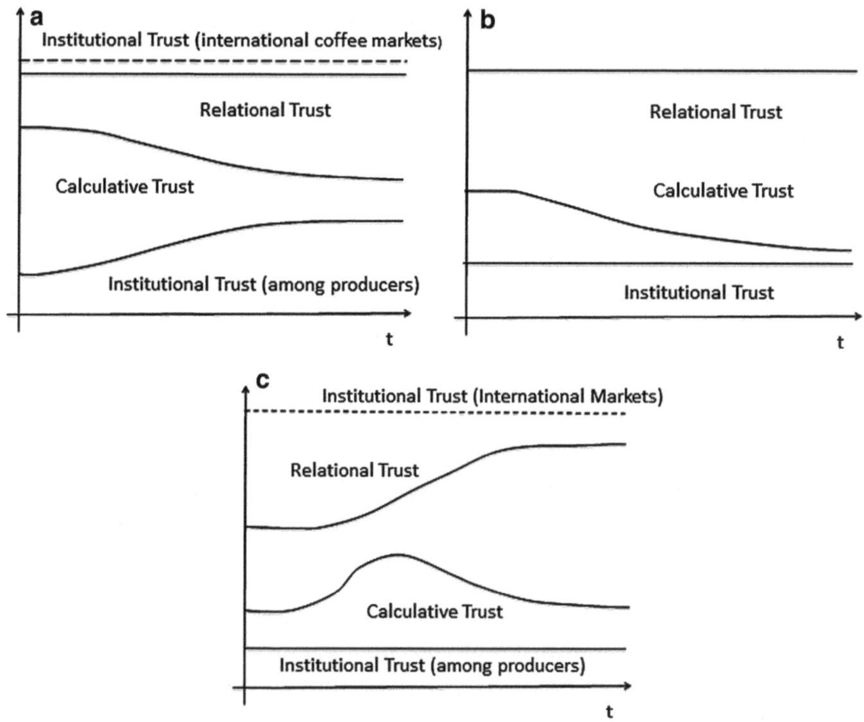

Fig. 3.1 Trust patterns over time. (**a**) Tosepan Titataniske. (**b**) Bats'il Maya. (**c**) Nestlé

trust the Nestlé brand name as an institutional anchor is an important driver. However, institutional trust of the Nestlé brand name as an institutional force among the producers is a much less important force. Rather, in the beginning more and more of the producers calculate that it may be in their interest to cooperate with Nestlé. Over time and as the case develops, some of this calculative trust converts to a more relational trust, making it less necessary for the producers to continually calculate the costs and benefits of their relationship with Nestlé.

Taken together the three cases depict trust as a complicated multidimensional concept that evolves dynamically over time. Based on the limited data from these three cases, we might expect the more calculative model of trust to be less permanent over time, while the relational model is more permanent but less generalizable. One weakness of the institutional model is long-term versus short-term calculation and how well the institution can adapt to changes in the market. A countervailing strength is that institutional trust is perhaps the best way to communicate with consumers in an international marketplace (Delhey & Newton, 2005).

3.6 Trust as a Design Characteristic in Large, Scalable FIPP Systems

In our larger research project, we were interested in how using advanced technologies such as cloud computing, interoperable data architectures, and social computing systems can support the development of scalable systems that approximate some of the desirable properties of the smaller-scale networks that we studied in the three cases. We expect that all three forms of trust—institutional, calculative, and relational—will play a large role in all forms of FIPP customer-oriented systems and networks. However, both the literature and our cases suggest that relational and institutional trust may play a more central role compared to calculative trust (Gefen et al., 2006; Pavlou & Dimoka, 2006).

It seems clear to us that addingsocial computing components to such systems can be a good way to build relational trust by empowering consumer activists to play a more prominent role in FIPP-oriented markets. Research on trust in information systems indicates that most individuals trust peers over certification or branding (Pavlou & Dimoka, 2006). For example, the US Center for Disease Control (CDC) reports that most individuals take to heart its advice on a communicable disease such as H1N1 flu more readily when information is forwarded to them from a peer-computing site rather that when downloaded from an official CDC site (Nall, 2010). But what information can social computing sites use to develop reliable ratings? Peer ratings based on consumer experiences with the product alone will not work since FIPP packages, by definition, report on unobservable attributes of products (e.g., when produced, by whom, and under what conditions).

Institutional trust will also play an important role (Gefen et al., 2006). However, simple and straight-forward certification systems will probably not work because as we will discuss in other chapters in the book, the explosive increase in certification systems makes it hard for consumers and other members of the supply chain to understand the meaning of each of these certifications. The answer may rest in some combination of system features such as providing online confirmable metadata about certifiers, providing information about how certifiers are certified, by allowing consumer advocates to rate certifiers as well as producers and supply chain operators, and perhaps even providing a legal status for some portion of the FIPP information package (with stiff penalties for providing false information or misusing such information as in the case of SEC prosecutions for insider trading).

In order to create information systems that can create such complex mixtures of trust, the missing key technical mechanism is a combination of data standards and procedures that allow data to be shared among user communities or a "data infrastructure building block."[4] In our research, we have been working on the creation of

[4] The name "data infrastructure building block" derives from National Science Foundation Data Infrastructure Building Block program which aims to "foster cross-community infrastructure development that solves common problems, while building blocks of data infrastructure that can support and provide data solutions to a broader range of scientific disciplines while reducing duplicative efforts." (http://www.nsf.gov/funding/pgm_summ.jsp?pims_id=504776)

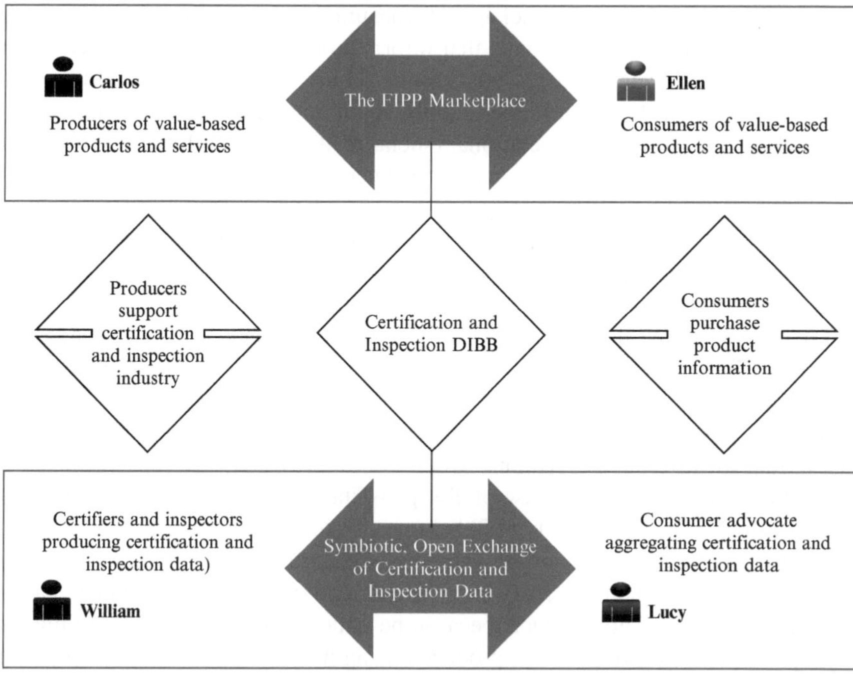

Fig. 3.2 Certification and Inspection Data Infrastructure Building Block (CIDIBB)

a proof-of-concept Certification and Inspection Data Infrastructure Building Block (CIDIBB). A CIDIBB is a set of data standards specifically tuned to the certification and inspection process that would allow certifiers and inspectors to post their data in a standard way and allow consumer advocates to gain structured access to the full information package that rests behind each certificate. Consumer advocates would know who did what, when, where, and under whose authority to create a given certificate. In Chapter 5, we provide a technical description of the CIDIBB prototype.

At a high level, Fig. 3.2 shows how CIDIBB could work. Each of the four classes of key stakeholders is represented by an idealized individual shown in each of the four corners.

Ellen, shown in the upper right-hand quadrant, represents consumers who will scan a product bar code to view its consumer rating to help them make a purchasing decision. Ellen is the end customer who is willing to pay a premium for a product if she receives information that she can trust indicating that the product has been produced in a manner that is consistent with her values as a consumer. She is the consumer who purchases certified coffee from Tosepan Titataniske, who goes out of her way to take her coffee break at the Azúcar Morena specialty coffee shop, and who recognizes the Nestlé brand name.

Lucy, in the lower right-hand quadrant, represents the emerging open data industry, where consumer advocates, among other different types of entrepreneurs, analyze the full information package of consumer products and then—through innovative business models—make that information available to consumers such as Ellen. Lucy's financial success rests on her ability to create a trusted business model appealing to Ellen and other supply chain actors, using a combination of calculative, institutional, and relational (social media) trust-building technologies and approaches. In current large-scale consumer markets, Lucy's role is not well developed. It might be best represented by consumer advisory services such as *GoodGuide*[5] or perhaps the popular version of *Consumer Reports*.[6] In our cases, Lucy might best be represented by the taste testers who work for at Bats'il Maya and promote high-quality coffee in Azúcar Morena.

Lucy relies on William, a member of the inspection and certification industry, who uses CIDIBB to broadcast information about how, when, where, and by whom consumer products are created. The marketplace will drive what certificates William is creating depending on what issues consumers are concerned with. In our three cases, the ideal type of William would be represented by Certimex, Ocia International, and FLO in the case of Tosepan Titataniske or perhaps the Common Code for the Coffee Community (4C) being promoted by the Nestlé Corporation.

Carlos represents producers of value-based products and services. Carlos is cooperating with William to certify his production processes and to document unobservable attributes of his products because he understands that Ellen is willing to pay a price premium for products produced using methods that are congruent with her values. In order to stay in business, both William and Carlos will need to sell trusted information to Lucy based on institutional assurances (e.g., Fair Trade, USDA Organic, and other virtual certifications that may come into existence). In our case studies, the ideal-type producer of "Carlos" is represented by the small coffee producers in the Tosepan Titataniske cooperative, those cooperating with Bats'il Maya as well as the 80,000 producers who register with Nestlé's 4C production registry. In the future, as large-scale Internet-based systems based on CIDIBB emerge, these systems will possess an ability to "drill down" and test the logic, provenance, and trustworthiness of that information using sophisticated probes and queries that Lucy may deploy to search William's data (an extreme version of calculative trust-building mechanisms) based on data provided by Carlos in the first instance.

Our previous research has shown that trust plays a key role in all aspects of information transfer around a distributed CIDIBB structure such as that sketched in Fig. 3.2 (Luna-Reyes et al., 2013). In fact, trust is considered as an alternative governance mechanism in most collaborative relations (Powell, 1996). Higher trust levels lead to lower costs resulting from the need to protect against opportunism (Shapiro, Sheppard, & Cheraskin, 1992). Moreover, the literature points out the importance of trust in these market transactions, particularly in the case of unobservable product attributes (Arora, 2006).

---

[5] http://www.goodguide.com

[6] http://www.consumerreports.org

3.7 Concluding Remarks

Institutional factors, based largely on certification and traceability systems, create an environment conducive to collaboration among producers and distributors, as well as channels for trustworthiness to be forwarded from producers over the supply chain to the customers, thus reducing the risks of information asymmetry. Our cases show little relevant evidence about the ways in which calculative trust contributes to FIPP networks, which suggests that its importance is limited, and may only be useful in establishing initial relationships between producers and distributors. Since consumers do not often engage in transactions directly with producers—at least in our three cases—calculation of risks and gains may not be a realistic action to take.

The role of relational trust is significant for both the relationship among producers and distributors and the relationship between producer and consumers in all of the cases that we reported. In all cases, deference based on reciprocal interactions and shared identities are critical for actors to share knowledge and information, build relatively permanent reliance on each other, and coordinate production and distribution activities. In addition, we can observe that the producers are also trying to reach end consumers and bridge the connection through identity alignment, such as producer-consumer connections built in the network of cafes in Bats'il Maya.

Another observation we had is that trust is not static. It evolves over time during which time the three forms of trusts interact. For example, institutional trust may ease the way for the development of calculative trust and relational trust. In FIPP networks, because of the information asymmetry problem, it is difficult for consumers to calculate risks and differentiate qualities. In our cases, certification and traceability systems (sources of institutional trust) appear to have an important role in building new relationships with producers and distributors. In the case of the Tosepan Titataniske cooperative, for example, fair-trade and organic certification labels established an ability to penetrate export markets reaching a different class of consumers. The result is that the evaluation of trustworthiness of information provided to consumers can be more effective which in turn drove the development of shared values between producers in Mexico and consumers in the North American coffee market. Similar mechanisms are also observed in the other cases.

The literature suggests that calculative trust can be turned into relational trust after a certain period of positive transactions. We do see some evidence for this relationship in the Nestlé case. Over time, conscious calculation of a coffee producer can be replaced with voluntary association with a cooperative in his or her community, sharing the same set of principles and values. This phenomenon could be especially apparent in FIPP networks as it promotes certain social and environmental objectives at the community level.

What seemed to be different across the cases, however, is that trust development may not take a single path. In the cases of the Tosepan cooperative and Bats'il Maya, high levels of relational trust appear early and dominate commercial relationships. In the Nestle case, relational trust, when it comes at all, seems to come later and is predicated by early calculations of costs and benefits. In both the

Tosepan Titataniske and Nestle cases, institutional trust at the level of international coffee markets came first, either in the form of third-party certification such a FLO labeling or an information system providing reliable information about the products such as Nestlé's 4C system. However, institutional trust is not always a sufficient condition to succeed. In the case of Tosepan, for example, relational trust has played an important role in taking advantage of certification in exporting their coffee.

In order for the FIPP production to survive, FIPP producers need to prove to end consumers that their products are handled in certain ways that meet consumers' needs and deliver more social and economic value than standard products. In this way, what really differentiates FIPP products from standard products is the information associated with the production and the trustworthiness of sources of information. As a result, trust is of fundamental importance for building the confidence of consumers, mitigating the risks associated with information asymmetry and, ultimately, driving the demand for FIPP production. In a way, trust is the essential lubricant for two major types of relationships in FIPP networks. One is the relationship among producers and distributors, and the other is the direct relationship between producers and consumers.

In sum, these case studies provide helpful clues that will point the way to future system designers who aspire to create scalable FIPP systems by designing the attributes and features that such high-tech systems must have to attain the high-quality and trusted performance of their small-scale, often face-to-face predecessors.

References

Akerlof, G. A. (1970). The market for "lemons": Quality uncertainty and the market mechanism. *The Quarterly Journal of Economics, 84*(3), 488–500.

Albersmeier, F., Schulze, H., Jahn, G., & Spiller, A. (2009). The reliability of third-party certification in the food chain: From checklists to risk-oriented auditing. *Food Control, 20*(10), 927–935. http://doi.org/10.1016/j.foodcont.2009.01.010.

Arora, R. (2006). Product positioning based on search, experience and credence attributes using conjoint analysis. *Journal of Product & Brand Management, 15*(5), 285–292. http://doi.org/10.1108/10610420610685695.

Backhouse, J., Hsu, C., Tseng, J. C., & Baptista, J. (2005). A question of trust. *Communications of the ACM, 48*(9), 87–91. http://doi.org/10.1145/1081992.1081994.

Ba, S., & Pavlou, P. A. (2002). Evidence of the effect of trust building technology in electronic markets: Price premiums and buyer behavior. *MIS Quarterly, 26*(3), 243. http://doi.org/10.2307/4132332.

Burkert, M., Ivens, B. S., & Shan, J. (2012). Governance mechanisms in domestic and international buyer–supplier relationships: An empirical study. *Industrial Marketing Management, 41*(3), 544–556. http://doi.org/10.1016/j.indmarman.2011.06.019.

Delhey, J., & Newton, K. (2005). Predicting cross-national levels of social trust: Global pattern or Nordic exceptionalism? *European Sociological Review, 21*(4), 311–327. http://doi.org/10.1093/esr/jci022.

Eisenhardt, K. M. (1989). Building theories from case study research. *The Academy of Management Review, 14*(4), 532–550.

Eisenhardt, K. M., & Graebner, M. E. (2007). Theory building from cases: Opportunities and challenges. *Academy of Management Journal, 50*(1), 25–32. http://doi.org/10.5465/AMJ.2007.24160888.

Gefen, D., Pavlou, P., Benbasat, I., McKnight, H., & Stewart, K. (2006). ICIS panel summary: Should institutional trust matter in information systems research? *Communications of the AIS, 17*(Article 9), 205–222.

Jahn, G., Schramm, M., & Spiller, A. (2005). The reliability of certification: Quality labels as a consumer policy tool. *Journal of Consumer Policy, 28*(1), 53–73. http://doi.org/10.1007/s10603-004-7298-6.

Luna-Reyes, L. F., Zhang, J., Roy, R., Andersen, D. F., Whitmore, A., & Andersen, D. L. (2013). Information strategies to support full information product pricing: The role of trust. *Information Polity, 18*(1), 75–91.

Nall, J. (2010). CDC 2.0: Using social media to improve public health. In *Proceedings of the 11th Annual International Digital Government Research Conference on Public Administration Online: Challenges and Opportunities, Digital Government Society of North America*. Puebla, Mexico: ACM.

Pan, Y. (2011). Eliminating the cyber "lemons" problem with the e-reputation in e-commerce market: Theoretical model and practice. *International Journal of Networking and Virtual Organisations, 8*(3/4), 182. http://doi.org/10.1504/IJNVO.2011.039993.

Pavlou, P. A., & Dimoka, A. (2006). The nature and role of feedback text comments in online marketplaces: Implications for trust building, price premiums, and seller differentiation. *Information Systems Research, 17*(4), 392–414. http://doi.org/10.1287/isre.1060.0106.

Pavlou, P. A., Liang, H., & Xue, Y. (2007). Understanding and mitigating uncertainty in online exchange relationships: A principal- agent perspective. *MIS Quarterly, 31*(1), 105–136.

Powell, W. W. (1996). Trust-based forms of governance. In *Trust in organizations: Frontiers of theory and research* (pp. 51–67). Thousand Oaks, CA: SAGE Publications, Inc. Retrieved from http://knowledge.sagepub.com/view/trust-in-organizations/SAGE.xml

Román, S. (2010). Relational consequences of perceived deception in online shopping: The moderating roles of type of product, consumer's attitude toward the internet and consumer's demographics. *Journal of Business Ethics, 95*(3), 373–391. http://doi.org/10.1007/s10551-010-0365-9.

Rousseau, D. M., Sitkin, S. B., Burt, R. S., & Camerer, C. (1998). Not so different after all: A cross-discipline view of trust. *Academy of Management Review, 23*(3), 393–404. http://doi.org/10.5465/AMR.1998.926617.

Saini, A. (2010). Purchasing ethics and inter-organizational buyer–supplier relational determinants: A conceptual framework. *Journal of Business Ethics, 95*(3), 439–455. http://doi.org/10.1007/s10551-010-0432-2.

Saparito, P. A., Chen, C. C., & Sapienza, H. J. (2004). The role of relational trust in bank – Small firm relationships. *Academy of Management Journal, 47*(3), 400–410. http://doi.org/10.2307/20159589.

Shapiro, D. L., Sheppard, B. H., & Cheraskin, L. (1992). Business on a handshake. *Negotiation Journal, 8*(4), 365–377. http://doi.org/10.1111/j.1571-9979.1992.tb00679.x.

Sheppard, B. H., & Sherman, D. M. (1998). The grammars of trust: A model and general implications. *Academy of Management Review, 23*(3), 422–437. http://doi.org/10.5465/AMR.1998.926619.

Smith, S., & Barrientos, S. (2005). Fair trade and ethical trade: Are there moves towards convergence? *Sustainable Development, 13*(3), 190–198. http://doi.org/10.1002/sd.277.

Sriram, M. S. (2005). Information asymmetry and trust: A framework for studying microfinance in India. *The Journal for Decision Makers, 30*(4), 77–85.

Stake, R. E. (1995). *The art of case study research*. Thousand Oaks, CA: Sage Publications.

Thornton, L. M., Esper, T. L., & Morris, M. L. (2013). Exploring the impact of supply chain counterproductive work behaviors on supply chain relationships. *International Journal of Physical Distribution and Logistics Management, 43*(9), 786–804. http://doi.org/10.1108/IJPDLM-09-2012-0298.

Viitaharju, L., & Lähdesmäki, M. (2012). Antecedents of trust in asymmetrical business relationships: Differing perceptions between food producers and retailers. *Marketing Intelligence & Planning, 30*(5), 567–587. http://doi.org/10.1108/02634501211251061.

White-Cooper, S., Dawkins, N. U., & Anderson, L. A. (2009). Community-institutional partnerships: Understanding trust among partners. *Health Education & Behavior, 36*(2), 334–347.

Yin, R. (1994). *Case study research: Design and methods*. Newbury Park, CA: Sage.

Zornoza, A., Orengo, V., & Penarroja, V. (2009). Relational capital in virtual teams: The role played by trust. *Social Science Information, 48*(2), 257–281. http://doi.org/10.1177/0539018409102414.

Chapter 4
Labeling, Certification, and Consumer Trust

Djoko S. Sayogo, Holly Jarman, David F. Andersen, and Joanne S. Luciano

Abstract Increased interest in ethical consumption has promoted the creation of incentives for product differentiation, which has been adopted by the market in terms of a variety of labels and certificates to describe a whole collection of product attributes related to health, social, or environmental sustainability. In this chapter, we describe and compare six coffee certifications in terms of their certification processes, governance mechanisms, and market penetration. Our comparison shows that leading certifications reassert their trustworthiness by emphasizing transparency, legitimacy, and accountability of their practices and governance processes. To demonstrate transparency, it is common that certification authorities openly publicize their standards and principles to demonstrate the transparency. To show legitimacy, they get accreditations from reputable national or international organization. Unfortunately, most of this information is not always at the reach of final consumers.

Keywords Certification • Labeling • Governance mechanisms • Greenwashing • Bluewashing • Consumer trust

---

D.S. Sayogo (✉)
University of Muhammadiyah at Malang, Jl. Raya Tlogomas no. 246, Malang, Indonesia

Center for Technology in Government, University at Albany
187 Wolf Rd., Suite 301, Albany, NY 12205, USA
e-mail: dsayogo@ctg.albany.edu

H. Jarman
Department of Health Management and Policy, School of Public Health,
University of Michigan, Ann Arbor, MI, USA
e-mail: hjarman@umich.edu

D.F. Andersen
Rockefeller College of Public Affairs and Policy, University at Albany, Albany, NY, USA
e-mail: david.andersen@albany.edu

J.S. Luciano
Tetherless World Constellation, Rensselaer Polytechnic Institute, Troy, NY, USA
e-mail: jluciano@gmail.com

H. Jarman, L.F. Luna-Reyes (eds.), *Private Data and Public Value*,
Public Administration and Information Technology 26,
DOI 10.1007/978-3-319-27823-0_4

4.1 Introduction

A number of studies of ethical and sustainable consumption suggest that consumer trust in ethical products is engendered partially based on the company's ethical conduct and the ethical and sustainable label attached to the product (see, e.g., Carrigan & Attalla (2001), Castaldo, Perrini, Misani, and Tencati (2009), Janssen and Hamm (2011), Pivato, Misani, and Tencati (2008), Polonsky, Bhaskaran, and Cary (2005); Swaen and Chumpitaz (2008)). However, the current rapid growth of certification and labeling schemes (Bacon, 2005; Muradian & Pelupessy, 2005; Raynolds, Murray, & Heller, 2007)[1] makes critics question the extent to which a particular ecolabel truly reflects the sustainability of the product and whether labels and certifications have simply become marketing gimmicks for large corporations (Gibson, 1999).

This chapter explores the ways existing certifications and labels enhance consumer trust in a particular product. We examine six coffee certifications to assess the adequacy of private regulation as a trust-inducing tool. We chose sustainable certified coffee as our case study for three reasons. First, sustainable certified coffee provides comprehensive yet manageable overview of the complexities in sustainable certification and labeling practices. Second, certification of coffee represents one of the most rapidly growing areas of certification (Raynolds et al., 2007). Third, coffee is one of the most traded commodities in the world market (Taylor, Murray, & Raynolds, 2005).

We examine the certifications in respect to their certification process, governance mechanism, and market growth. Through this comparison, we found certification schemes that use different strategies to reassert the trustworthiness of their certificate to the consumer and general public. However, we also found that information about the steps certifications are taking to increase their trustworthiness is not readily apparent or available to consumers.

We begin this chapter with a look at the relationship between government intervention, private regulation, and market governance. We follow with an examination of the utility of certifications and labels in helping consumers make informed decisions and describe some of the shortcomings of the current certification system. The subsequent sections, section four to six, focus on examining six different labeling and certification schemes. Section four presents the coverage and scope of certifications and labels in terms of operational scope, market growth, market penetration, and growth strategy. Section five focuses on the governance process and section six outlines the differences in certification assessment processes. Finally, section seven provides the concluding remarks.

[1] There is a rapid and sustained growth for certified products, especially food products with coffee in particular (Raynolds et al., 2007), especially after the coffee crisis (Muradian & Pelupessy, 2005). There is also rapid growth of ecolabels in general; ecolabel index in 2012 tracks the existence of 435 ecolabels worldwide (www.ecolabelindex.com).

4.2 Government Interventions, Private Regulation, and Market Governance

Smart disclosure is seen as one way of creating market intervention by allowing consumers make better purchasing decisions. Information can be used to create market interventions directly and indirectly (Weiss, 2002). Direct intervention occurs when the government collects and distributes information directly to the public. An example of such direct intervention is the publication of information on chemical and toxic substances manufactured in, or imported into, the United States[2] by the Environmental Protection Agency (EPA). Indirect intervention occurs when nongovernment actors generate or share information that is required or enabled by the government (Weiss, 2002). Product labeling is an example of an indirect intervention and this chapter will focus on this type of intervention.

Indirect interventions through product labeling can either be mandated by government or be voluntary in nature. Mandatory policies for product labeling are backed by government regulation and require compliance by all market actors without any exceptions. For example, the US Nutritional Labeling and Educational Act of 1990[3] requires all manufacturers to attach nutrition labels to their products, and the Alcohol Beverage Labeling Act of 1988 requires two mandatory warnings to be placed on all alcoholic beverage containers.

In contrast, voluntary requirements for product labeling allow market actors to adopt or ignore measures as they see fit. Most labeling and certification schemes adopt a voluntary approach whereby market actors complying with a set of standards may attach a label onto their product based on their own interests. Producers use labels to maintain and enhance their reputation, boost consumer trust in their products, and differentiate themselves from their competitors in the marketplace. For example, companies use labels and certifications to support the credibility of their claims on issues such as environmental sustainability or human rights (Fig. 4.1).

Voluntary policies exist under two different types of governance regime, hybrid or market based. Under hybrid systems of governance, voluntary standards can be created or enabled by the government, but are administered by an independent body. For example, the government of Quebec created an independent organization to monitor labeling systems regarding the origin and authenticity of products sold within the province, called CARTV.[4] Similarly, the US Department of Agriculture (USDA) enacted National Organic Program (NOP) to regulate production and handling of organically produced agricultural products. The standards are created by USDA but the certifying process is conducted by third-party certifying agents accredited by the USDA NOP.[5]

---

[2] http://java.epa.gov/oppt_chemical_search

[3] http://www.fda.gov/ICECI/Inspections/InspectionGuides/ucm074948.htm

[4] *Conseil des appellations réservées et des termes valorisants (CARTV),* http://www.cartv.gouv.qc.ca/en/about-us

[5] http://www.usda.gov/wps/portal/usda/usdahome?navid=ORGANIC_CERTIFICATIO

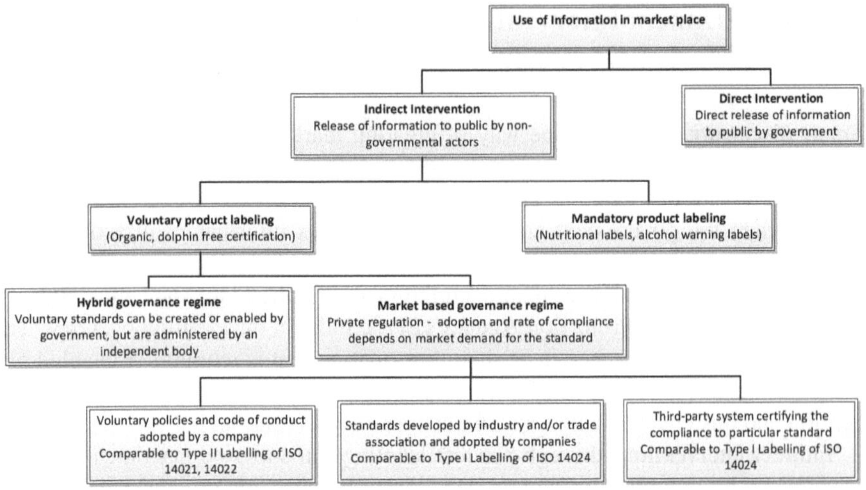

Fig. 4.1 Categories of product labeling

Under market governance regimes, standards are "market driven." This means that their adoption among organizations, and the rate of compliance, depends on market demand for the given standards (Cashore, Auld, & Newsom, 2004). Certification schemes under market governance are commonly referred to as private regulation. Private regulation is a governance system formed by coalition of nongovernment actors to codify and monitor conduct of private entities in respect to issues such as sustainable and ethical production (Bartley, 2007; Büthe, 2010a, 2010b; Mayer & Gereffi, 2010). Private regulation is designed to use market pressure to regulate the behavior of industry's actors (Bartley, 2007; Mayer & Gereffi, 2010) and to add layers to the existing laws, regulations, and standards enacted by the government (Bartley, 2011). Private regulation takes many forms (standards, codes of conduct, and certification systems) and is organized in three different formats: (a) privately developed voluntary policies and codes of conduct, (b) standards developed by industry and/or trade association and adopted by companies, and (c) third-party certification systems in which independent monitors certify company's voluntary compliance to a particular standard (Bartley, 2007; Mayer & Gereffi, 2010).

Third-party certification systems are deemed more trustworthy than the other two forms of private regulation because of the independent nature of monitoring, which eliminates potential conflicts of interest (Jahn, Schramm, & Spiller, 2005). By providing trustworthy information, third-party certifications and labels help alleviate information asymmetry between consumers and producers. In asymmetric relationship, consumers have limited access to information that would help them accurately assess invisible product attributes such as safety, quality, or social and environmental sustainability. Third-party certifications provide assurances to consumers regarding the credibility of product attributes as represented by the product label. This assurance is especially important in reducing information asymmetry regarding product attributes related to internal production methods that are difficult and/or economically infeasible for consumers to access and evaluate by themselves. Some examples of these attributes are the robustness of automobile engine quality, toothpaste's ability to reduce plaque, or a claim that a product produced under a fair-trade standard improved living conditions of small farmers or plantation workers.

4.3 The Utility of Labels in Assisting Consumer Choice

Consumer's purchasing decisions are influenced by competing priorities (Szmigin, Carrigan, & McEachern, 2009), which can create a gap between professed intention to purchase an ethically produced product and the actual purchase of such product (Carrigan & Attalla, 2001; Carrington, Neville, & Whitwell, 2010). Despite this gap, consumers recognize the utility of certifications and labels using them as a substitute for searching for more information (Carrigan & Attalla, 2001) and as trust-inducing tools (Carrigan & Attalla, 2001; Janssen & Hamm, 2011; Polonsky et al., 2005). For instance, consumers positively correlate organic certification with greater level of trust in a given product (Janssen & Hamm, 2011).

The usefulness of labels and certifications depends on the extent to which consumers understand the information behind them (Carrigan & Attalla, 2001). A number of studies found that consumers' understanding of what a particular label or certification conveys is rather limited (Carrigan & Attalla, 2001; Janssen & Hamm, 2011; Polonsky et al., 2005). One of the reasons for the limited understanding is the rapid proliferation and increasing diversity of third-party certifications and labels.[6] The growing complexity of the third-party certification environment increases information processing demands on consumers and consequently diminishes the meaning of a certification as a trust element. In addition, the large number of certification and labeling schemes complicates efforts to assess and compare the credibility and quality of labels (Jahn et al., 2005; Raynolds et al., 2007), which in turn increases consumers' need for additional information (Pelsmacker, Janssens, Sterckx, & Mielants, 2005).

The remainder of this chapter provides a comparison of existing certifications to illustrate the strengths and weaknesses of certifications as a trust-inducing tool. We compare and contrast six major coffee certification schemes[7]: Fairtrade International (FLO), Rainforest Alliance Network (RAN), UTZ Good Inside, Common Code for the Coffee Community (4C), the US Department of Agriculture (USDA) National Organic Program, and Coffee and Farmer Equity (C.A.F.E.) Practices. We exclude Nespresso AAA due to its exclusive focus on coffee quality and less on social and environmental sustainability. The comparison addresses three aspects: the scope of the certification, the governance of third-party certifiers, and the certification and inspection processes.

[6] Data from EcoLabelIndex shows the existence of 435 ecolabels worldwide (Ecolabelindex, n.d.).

[7] Coffee Barometer 2012 indicates the existence of seven major coffee initiatives, namely, FLO, UTZ, 4C, Organic, RAN, C.A.F.E. Practice, and Nespresso AAA (Panhuysen & van Reenen, 2012).

4.4 The Scope of Certifications and Labels

There are two different scopes we have to consider when talking about certifications: the operational scope and the market scope. The operational scope refers to the operational capacity of a certification such as year of establishment, focus area, and legal status. Market scope refers to the economic growth of the certification in term of market share and penetration.

4.4.1 The Operational Scope of Certifications and Label Initiatives

Operational scope of a certification, which includes its year of establishment, legal status, and area of focus, can impact the level of trust a consumer has in that particular certification. Three certifications (FLO, RAN, and USDA Organic) were founded in the 1990s, while the other three certifications (UTZ, 4C, and C.A.F.E.) were established in the 2000s.[8] A study by Rao (1994) found a significant relationship between organization's age and survival of that organization, which implies that newer certifications are more vulnerable to competition because of lower brand recognition. Similarly, age of certification might correlate with its reputation. Thus consumers might assume that older certifications are more trustworthy because they have better public recognition.

Certifications in the certified sustainable coffee context differentiate themselves by different areas of focus, either concentrating on social impact, environmental impact, or a combination of both. For instance, USDA Organic only focuses on environmental aspects, while FLO focuses on both social and environmental values. Each of these areas can include different specializations, ranging from child labor to health to safety of work environment. Some certifications, such as FLO and RAN, add distinct focus areas to differentiate themselves. FLO adds gender issues, while RAN adds local communities (Table 4.1).

Objectiveness of the evaluation process is crucial for the reliability of a certification (Deaton, 2004; Tanner, 2000), which can be achieved by having certification processes accredited by an independent organization that has higher authority than the certifiers (Deaton, 2004). Being accredited is crucial to soliciting trust because it augments the reputation and legitimacy of the certifiers. As presented in Table 4.1, four certification schemes, FLO, UTZ, 4C, and RAN, are accredited by ISEAL.[9] USDA Organic, on the other hand, uses its status as a government agency to act as a standard setter as well as an accreditation body. C.A.F.E. Practices was established and is vouched for by a private company (Starbucks) and has no affiliation with any accreditation body.

---

[8] http://www.ecolabelindex.com/ecolabel/cafe-practices

[9] ISEAL is a global membership association in which sustainable standards could be admitted if they meet ISEAL Code of Good Practice (http://www.isealalliance.org/about-us).

Table 4.1 The operational scope of certifications and labels

Indicator	FLO	UTZ	4C	RAN	USDA	C.A.F.E.
Year of establishment	1993	2002	2006	1992	1990	2004
Social area covered	Child labor, employment practice, *gender issues*, health and safety work, ILO 8 core conventions, work, and labor rights	Child labor, employment practice, health and safety work, ILO 8 core conventions, work, and labor rights	Child labor, employment practice, health and safety work, ILO 8 core conventions, work, and labor rights	Child labor, employment practice, health and safety work, ILO 8 core conventions, *local communities*, work, and labor rights	Not applicable	Child labor, employment practice, health and safety work, ILO 8 core conventions, work, and labor rights
Status of certification	Private voluntary	Private voluntary	Private voluntary	Private voluntary	Public voluntary	Private voluntary
Accreditation	Full ISEAL member	Full ISEAL member	Full ISEAL member	Full ISEAL member (SAI)	Not applicable	Not applicable
Managing Organization	Fairtrade International e.V. (FLO)	UTZ Certified Foundation	4C Association	Sustainable Agriculture Network	US Department of Agriculture	Starbucks and Conservation International
Status of founder(s)	Majority nonprofit entities	Nonprofit and profit entities	Nonprofit and profit entities	Majority nonprofit entities	Government	Private entities

Source: Majority of the data was obtained from the voluntary standard analysis and research of the ITC (International Trade Center) (http://search.standardsmap.org) and data for C.A.F.E. was extracted from the Conservation International website (http://www.conservation.org/campaigns/starbucks/Pages/CAFE_Practices_Results.aspx)

4.4.2 The Market Growth and Coverage of Certifications

One key to mainstreaming ethical and sustainable consumption is rooted in the assumption that an increase in demand for sustainable products sends signals to the rest of actors in the supply chain to conform to sustainability requirements (Seyfang, 2005). Consequently, the market growth for sustainable product signifies the power of consumers to change the behavior of supply chain actors. Consumers could use their purchasing behavior to exit their relationship with the company (Hirschman, 1970) or to express their political or ethical values to the company (Howard & Allen, 2010). Different certifications/labels receive different reception in different markets (Pierrot, Giovannucci, & Kasterine, 2011).

- *Fair Trade (FLO) Certified Coffee:* Fair Trade certification is the second largest type of certification for coffee produced and consumed worldwide (Raynolds et al., 2007). FLO certified coffee is very dominant in the United Kingdom, France, and recently in the United States. The 2011 annual report from FLO indicates that the worldwide volume of sales for fair-trade organic and fair-trade conventional grew by 6.7 % in just 3 years, from 12 % in 2008 to 2009 to 18.70 % from 2009 to 2010. In the United States, the growth rate of Fair Trade certified coffee experienced dramatic increase of 32 % from 2010 to 2011[10] and the US fair-trade market accounted for more than 10 % of total fair-trade sales worldwide. In 2011, the majority of Fair Trade certified coffee imported to the United States came from Latin America (86 %) with the rest coming from Asia (approx. 10 %) and Africa (4 %) (Fig. 4.2).
- *Rainforest Alliance (RAN) Certified Coffee:* RAN is regarded as the third largest initiative for NGO-based coffee certification after organic and fair trade (Raynolds et al., 2007). RAN is also the market leader in Japan (Pierrot et al., 2011). RAN press release claims that RAN certified coffee represented approximately 3.3% of the global coffee market share.[11] The growth rate of RAN certified coffee is impressive, with increase of 13 % in sales volume from 2010 to 2011, from approximately 114,924 metric tons in 2010 to 129,864 metric tons in 2011. The production of RAN certified coffee grew by 20 % from 2010 to 2011 (RAN, 2012). The impressive growth of RAN certified coffee was attributed to the increasing commitment from their alliances with mainstreams and large coffee companies, such as: Kraft Food, Nespresso, Tchibo, and others (Kolk, 2010; Pierrot et al., 2011; Raynolds et al., 2007). Nespresso, for example, committed to certifying 80 % of their coffee with the Rainforest Alliance by 2013 (Pierrot et al., 2011) (Fig. 4.3).
- *UTZ Certified Good Inside:* UTZ certified coffee achieved strong growth in the European market, particularly in the Netherlands (Pierrot et al., 2011). Approximately 30 % of coffee consumed in the Netherlands have UTZ label

[10] See the Coffee Almanac published by the Transfair USA for 2011 (http://www.fairtradeusa.org/sites/default/files/Almanac%202011.pdf).

[11] http://www.rainforest-alliance.org/newsroom/news/annual-growth-2011

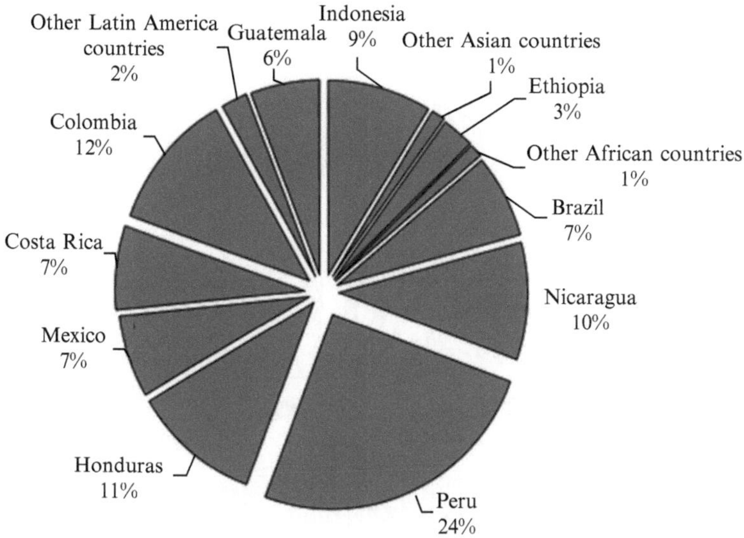

Fig. 4.2 Percentage of Fair Trade certified coffee imported into the United States by country of origin, 2011 (Source: Impact report fair trade USA[11])

Fig. 4.3 Worldwide imports of Rainforest Alliance certified coffee, 2009 (Source: adopted from Pierrot et al. (2011))

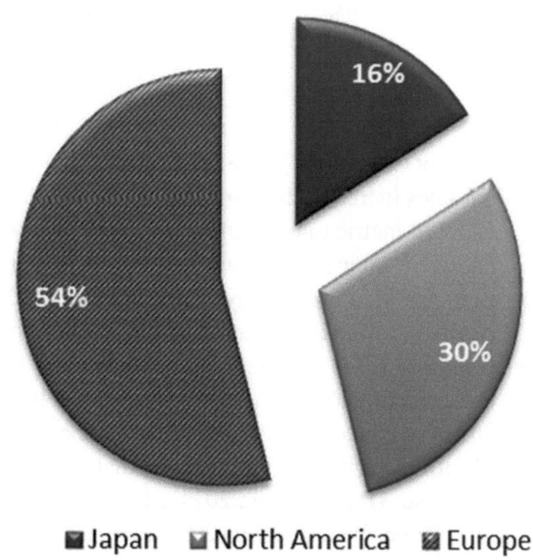

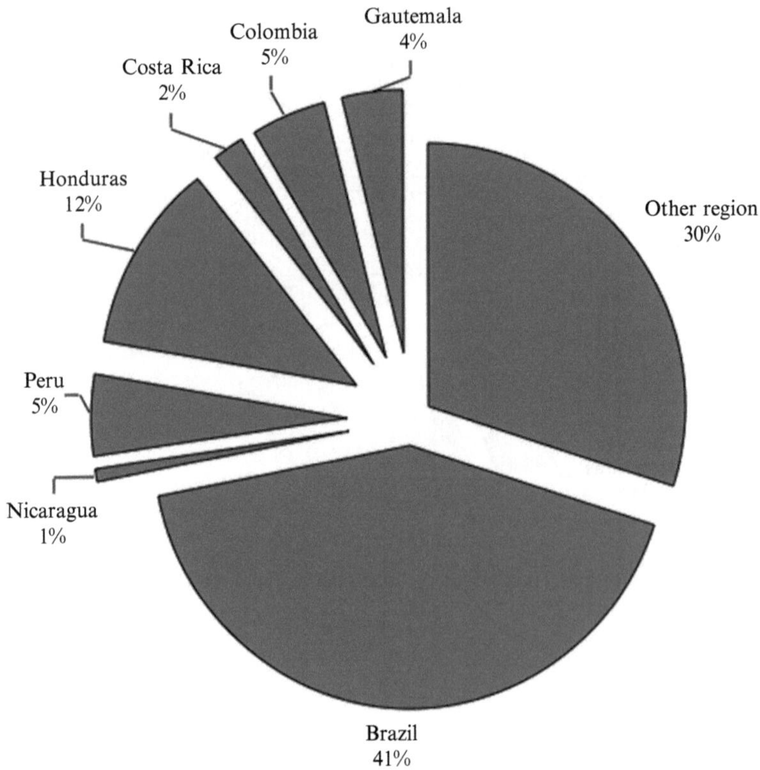

Fig. 4.4 UTZ certified coffee imported worldwide by country of origin, 2010 (Source: Supply demand update UTZ 2011[13])

(Pierrot et al., 2011). The UTZ Supply and Demand update for 2011 indicates that the sales increased by almost 50 % from 2009 to 2010 with sales in 2010 reaching 121,234 metric tons.[12] Similar to RAN, the growth of UTZ certified coffee is influenced by their alliance with large mainstream market corporations such as Sara Lee, Ahold, and Safeway (Raynolds et al., 2007). The majority of imported UTZ certified coffee originated from the Latin America region (70 %) (Fig. 4.4).

- *Common Code for the Coffee Community (4C):* 4C is different from the other three certifications in two ways: first, 4C is designed around a business-to-business concept, while the other three are more consumer oriented, and second, 4C offers verification procedures that are less rigorous than the other certification processes (Pierrot et al., 2011). 4C annual report in 2010 indicates that their sales in volume from 2008 to 2009 increase by approximately 140 %, from 11,900

[12] See http://www.UTZcertified-trainingcenter.com/home/images/documentos/general/supply__demand_report_2011_UTZ_certified.pdf or http://www.katocoffee.com/com/info/goodinside/supply__demand_report_201104.pdf

metric tons in 2008 to 28,600 metric tons in 2009. Much of the increases in purchasing volumes were attributable to the purchasing commitments of 4C buying members, such as Nestle, Kraft Foods, and Tchibo, which were members of the 4C steering committee until December 2006 (Kolk, 2010).

- *USDA Organic*: USDA National Organic Program acts as standard setter as well as an accreditor for certifying bodies that will certify compliance to USDA requirements on their behalf. In terms of market growth, a survey by USDA's Economic Research Service indicates a significant 4.86 % growth in demand for organic food from $3.6 billion in 1997 to $21.1 billion in 2008 (Dimitri & Oberholtzer, 2009). The North American organic coffee market for 2009 accounted for $1.4 billion (Pierrot et al., 2011), which is roughly 6 % of the overall organic food market in 2008. Approximately 89 million pounds of organic coffee was imported into the United States and Canada in 2008 which represents a 12 % increase from 2007 (Pierrot et al., 2011).
- *C.A.F.E. Practices*: C.A.F.E. Practices is a standard developed by Starbucks in collaboration with the Conservation International in 2004 to ensure that coffee sold to Starbucks met their environment and social criteria and financial viability (Semroc, Baer, Sonenshine, & Weikel, 2012). Starbucks C.A.F.E. Practices is the single largest sustainable coffee certification in the United States (Raynolds et al., 2007; Pierrot et al., 2011). Starbuck's purchase of C.A.F.E. Practices certified coffee increased significantly; from 2007 to 2008, it increased to 77–81 % in 2009 to 84 % in 2010 (Semroc et al., 2012) and reached 86 % in 2011. Starbucks projected to have 93 % (509 million pounds) of their coffee supplies either certified by third-party certifier or through C.A.F.E. Practices by 2012 and to have them 100 % certified by 2015.[13] With high volume of purchase from C.A.F.E. certified coffee, only small number of coffee was certified by third-party certification. For instance, only 8.1 % (44.4 million pounds) of Starbucks coffee was Fair Trade certified and only 1.6 % (8.7 million pounds) was certified organic in 2012[18].

The review of market penetration by the six major sustainable coffee certifications points to three significant issues. First, the demand for certified coffee for all six types of certification is growing rapidly. The significant increase in market share of certified coffee implies that consumers increasingly trust certification, which in turn appears to drive increased for certified coffee products. Second, much of this rapid growth is attributable to alliances with mainstream coffee purchasers. The growth in RAN, UTZ, and 4C is propelled by connections to, and commitments by, mainstream coffee purchasers. Third, market growth of third-party certified coffee is restricted by certification schemes propagated by private companies, such as Starbuck's C.A.F.E. Practices.

[13] http://www.conservation.org/campaigns/starbucks/Pages/CAFE_Practices_Results.aspx and http://www.starbucks.com/responsibility/sourcing/coffee

4.5 The Governance of Third-Party Certifiers

The main objective of third-party certifiers is to provide a consumer some degree of assurance in respect to the invisible attributes of a product. The main selling point of third-party certifiers is to invoke trust from consumers through appeals to values such as independence, objectivity, and transparency (Deaton, 2004; Hatanaka, Bain, & Busch, 2005; Tanner, 2000). The claim of independence from conflict of interest (Hatanaka et al., 2005; Tanner, 2000) and the democratic nature of the decision-making process of the third-party certifiers (Raynolds et al., 2007) become distinguishing factors among different certification schemes. This section compares the governance mechanisms of the coffee certifiers in respect to participation and independence. Participation refers to the engagement of stakeholders in the governance of the certification, such as public comments during standard development. Independence refers to freedom from conflict of interest in the certification process.

The comparison of the independence and participation aspects of each certification focuses on three indicators: (1) compliance to national/international regulations, norms, and conventions, (2) democratic standard-setting process, and (3) the engagement of NGOs as coordinating organizations. The summary of the comparison is provided in Table 4.2.

- *Fair Trade (FLO) Certified Coffee:* FLO is a certification initiative with the broadest and strongest NGO support (Raynolds et al., 2007). The governance of FLO follows democratic mechanisms where members and other stakeholders can contribute to the strategy and standard setting through a general assembly. FLO has 25 members classified under five types: nineteen National Fairtrade organizations, three producer networks, four fair-trade marketing organizations, two fair-trade applicant members, and Flo-Cert, an independent certification body for fair-trade global certification. All members and other stakeholders are given the opportunity to participate at three annual assemblies: the general assembly, labeling initiatives' assembly, and producer network assembly. Certification bodies under FLO, such as Flo-Cert, are accredited by the ISO/IEC 65 to assure their independence in making certification decision.
- *Rainforest Alliance (RAN) Certified Coffee:* RAN is also regarded as certification with strong engagement with NGOs through their alliance to Sustainable Agriculture Network (SAN) (Raynolds et al., 2007; Pierrot et al., 2011). As of recent, RAN has not yet complied with the ISO 65 requirements. However, RAN established a set of principles to govern the integrity of their practice, including principles of independence and participation. Among others, these principles are (see www.rainforest-alliance.org) (a) separating the entity responsible for certification from the entity responsible for receiving donations; (b) limiting the objectives of contributions to fees from certifications, sponsorship for public events, and funding for educational activities; (c) initiating and creating public consultation via stakeholders' outreach, local workshops, or direct contacts; and (d) forming International Standards Committee with membership from stakeholders to improve their standards.

Table 4.2 Comparison of governance process

Indicator	FLO	4C	UTZ	RAN	USDA	C.A.F.E.
Multi-stakeholder engagement in governance and standard setting	Yes	Yes	Yes	Yes		No
Involvement of private entity in standard setting	n/a	Yes	Yes	n/a	No	Yes
General assembly or supervisory board as the highest authority	Yes	Yes	Yes	No	No	No
Engagement of NGOs in the governance mechanism	Yes	Yes	Yes	Yes	No	Yes
Compliance to national/ international regulations, norms, and conventions	Critical	Critical	Critical	Short term	Not covered	n/a
Policy for complaints against standard-setting organizations is available	Yes	Yes	Yes	No	Yes	n/a
Policy for complaints against certification body (CB) is available	Yes	n/a	Submit to CB	n/a	Yes	Submit to CB

Source: Majority of the data was obtained from the voluntary standard analysis and research of the ITC (International Trade Center) (http://search.standardsmap.org) and data for C.A.F.E. was extracted from the Conservation International website (http://www.conservation.org/campaigns/starbucks/Pages/CAFE_Practices_Results.aspx)

- *UTZ Certified Good Inside:* UTZ Good Inside assigned monitoring activities to independent organizations external to UTZ and demanded that these external verifiers complied with the ISO/IEC 65. UTZ demonstrates a multi-stakeholder approach in their governance involving public and private entities, especially the supervisory board and standard committee. For example, Sara Lee and Ahold serve on the board and standard committee of UTZ. Their supervisory board consists of combination of representatives from coffee companies, NGOs, and producer cooperatives (Raynolds et al., 2007). The standard committee of UTZ consists of between 6 and 12 individuals, ranging from private entities, NGOs, and academics.[14]
- *Common Code for the Coffee Community (4C):* 4C Association claims to promote participatory decision-making process by including coffee producers, trade and industry, and civil society members[15] in their governance approach. These tripartite components form three separate chambers that have equal voices in the governing entity of 4C Association. The governance consists of five elements: the general assembly, the council, the executive board, the technical committee, and the mediation board. Similar to UTZ, 4C also employs independent external

[14] See http://www.UTZcertified.org/en/whoweare/standards-committee

[15] See http://www.4c-coffeeassociation.org/aboutus/our-governance.html

certifiers to conduct their verification process and requires these certifiers to conform to ISO/IEC Guide 65.

- *USDA Organic:* USDA Organic is a state-based certification, meaning that this certification is supported by government, in this case the US Department of Agriculture. As a consequence, compliance to national or international regulations, norms, and conventions in regard to certification processes is not applicable to the USDA Organic certification. USDA Organic is also unique because USDA acts as standard setter and accreditation body at the same time. As standard setter, USDA develops the National Organic Program (NOP), a federal regulatory framework governing organic food in the United States guided by and based on the Organic Food Production Act of 1990 (http://www.ams.usda.gov). In developing the NOP, USDA solicited input from their citizen advisory board and the general public. As an accreditation body, USDA accredits certification organizations to inspect products for compliance to NOP on behalf of USDA. USDA also employs policy for complaints both against the standard-setting organization (USDA) and the certification bodies accredited by USDA (http://www.intracen.org/) to ensure accountability and integrity of the standard.
- *C.A.F.E. Practices:* There is less information about the governance of C.A.F.E. Practices. C.A.F.E. Practices is a result of collaboration between a private entity, Starbucks, and a nongovernmental organization, Conservation International (CI). As such, the development of standards in C.A.F.E. Practices was led by these two organizations. On the other hand, to ensure the integrity and credibility of the certification process, the enforcement of C.A.F.E. Practices is conducted by a third-party certifier, the Scientific Certification System (SCS) Global Service. The SCS accredits third-party certifiers to perform the certification process for C.A.F.E. Practices on their behalf. According to the Starbucks' website as well as Conservation International's website, there is no policy for complaints against the standard-setting organization.[16] However, complaints can be submitted against the third-party certifiers and the SCS Global Service.

The comparison of the six major coffee certifications indicates that they assert their credibility and legitimacy in three ways (Table 4.2). First, they promote their good governance through the engagement of multiple stakeholders, particularly engaging civil society organizations such as NGOs. For some certification, this multi-stakeholder engagement includes the involvement of private sector organizations, such as Sara Lee for UTZ.

Second, they emphasize their legitimacy by complying with national and or international regulations, norms, and conventions. This is especially the case for non-state certification schemes. The state-based certification such as USDA Organic uses their adherence to legislation to demonstrate legitimacy. Third, these certification schemes highlight the availability of mechanisms to lodge complaints against the standard-setting organization and/or the certification body to ensure accountability and integrity of the standard.

[16] For more information, refer to http://www.scsglobalservices.com/starbucks-cafe-practices.

4.6 The Assessment Processes of Third-Party Certifications and Labels

The robustness of an assessment process, also called audit, determines the reliability of a certification and label quality (Albersmeier, Schulze, Jahn, & Spiller, 2009; Jahn et al., 2005). As a result, the quality of the assessment process becomes the determining factor of the information quality of a label (Jahn et al., 2005). Borrowing from the financial audit literature, two factors affect the quality of an audit: the competence and the independence of the auditor (DeAngelo, 1981; Duff, 2004). Competence refers to the ability of the auditor to conduct due diligence and a thorough assessment of the audit object (DeAngelo, 1981). Auditor competence is also influenced by applicant's perception of auditor independence and elimination of conflict of interest (Ammenberg, Wik, & Hjelm, 2001).

In general, an applicant has to undergo two types of audit—a desk and a field audit. The desk audit compares the requirements listed in the standard against documentation submitted by an applicant. The desk audit is useful for planning the scope and focus of the field audit. During field audit, the auditor assesses the degree of compliance by reviewing the current practices of the applicant against a set of decision criteria. These decision criteria, usually called compliance criteria or control criteria, represent a set of measureable control points derived from the certification standard. In some certification schemes, such as FLO, the result of the field audit will be sent to the certifiers for final certification decision. Other certification schemes, such as UTZ, rely on the judgment of the auditor to make the final certification decision (Fig. 4.5).

The strengths and weaknesses of the assessment processes for each of the six certification schemes are presented in the paragraphs bellows, with a summary of the comparison presented in Table 4.3.

- *Fairtrade International (FLO).* FLO established Flo-Cert as an independent unit under FLO to organize and coordinate their certification process. As a certification body, Flo-Cert is conforming to the specific quality requirements in the ISO 65. The audit begins with a desk audit performed by Flo-Cert, which is followed by a field audit performed by a local auditor. Local NGOs and civil society representatives are involved during the audit process as a demonstration of the value that FLO places on local knowledge. Following field audit, the auditor sends the evaluation results to Flo-Cert for a quality check by the responsible certification analyst and for the final certification decision. After initial audit, the certificate is issued for 6 years, but it is contingent on passing an annual audit. The robustness of the FLO certification is enhanced by their requirement to conduct full audit of all functions in the applicant's organization, in contrast to a sampling audit, which only assesses certain functions of the organization. In addition, Flo-Cert ensures competence of their auditors by requiring comprehensive training and by establishing a designated entity that is responsible for evaluating the auditor's work and skills.

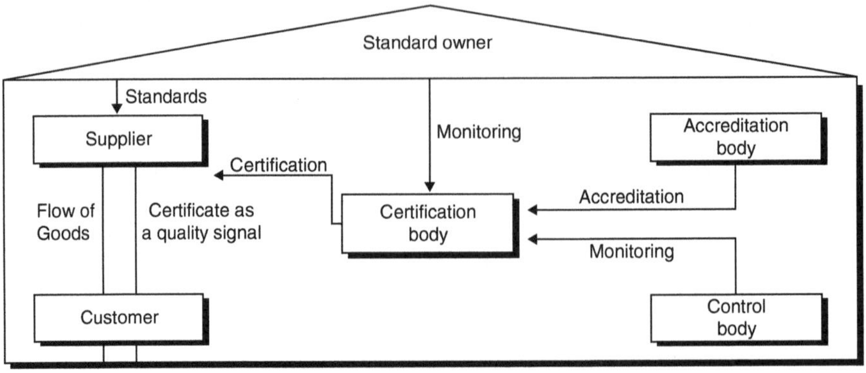

Fig. 4.5 The basic structure of certification process (Source: Jahn et al. (2005))

- *Rainforest Alliance Network (RAN)* uses internal auditors, which does not conform to the ISO 65 requirement.[17] RAN requests a written declaration of independence from the members who conduct audits to ensure independent nature of their assessments. RAN also makes sure that the member who conducts an audit is not involved in the final certification decision. The responsible authority for certification decision is the Sustainable Farm Certification International, Ltd (SAN). RAN also administers two types of audit, annual audit and surprise audit.[18] To ensure competence, RAN requires that their auditors acquire specific educational background, at minimum a university degree. The auditors are required to go through a training program to qualify for auditing responsibilities and are evaluated internally to assess their work and skills.
- *UTZ* relies completely on independent certification bodies external to UTZ for both the assessment process and certification decision. They require the independent certification bodies to be accredited by ISO and comply with ISO 65. In contrast to the other certification bodies, UTZ relies on the lead auditor for the final certification decision. UTZ consulted their stakeholders in the development of local indicator development, which is part of their code development procedure. UTZ also combines annual and surprise audit. Similar to the other certification schemes, UTZ requires the auditors to go through training program to ensure the competence of the auditor and the quality of audit. The rules and policies to evaluate the auditor's work and skills are developed by external certification bodies, but auditors are audited only if there is a complaint from the applicant.
- *The Common Code for Coffee Conducts (4C)* relies on external independent bodies, such as Control Union or Bio Cert, to perform their audits. In case of 4C, the final decision is still made by 4C and not the independent certifiers. These exter-

[17] Per July 2012, for more information, refer to http://www.rainforest-alliance.org/about/integrity/accountability

[18] Surprise audit means that the audit process is unplanned, unannounced, and without warning (Wells, 2002).

Table 4.3 Strictness of audit process

Indicator	FLO	4C	UTZ	RAN	USDA	C.A.F.E.
Involvement of certification body with governance mechanism of standard-setting organization	Yes	No	No	Yes	No	No
The use of local knowledge and stakeholder involvement	Yes	Yes	Yes	Yes	Yes	n/a
The use of local auditor to conduct the audit	Yes	n/a	n/a	No	Yes	Yes
The scope of the audit (full or sampling)	Full	No	n/a	No	n/a	n/a
The frequency of the audit	Annual	Surprise	Annual, surprise	Annual, surprise	Annual, validity 365 days	Annual, validity 365 days
Independency of auditor	Yes	Yes	Yes	Limited	Yes	Yes
Auditor undergoes training program to qualify for audit	Yes	Yes	Yes	Yes	Yes	Internal by the certifier
Information on certification decision is made accessible to the stakeholders	Public	Public	Public	n/a	Public	n/a
Certification body is independent of standard setter	Yes	Yes	Yes	No	Yes	Yes
Specific educational background is requested to act as auditor	No	No	Yes	Yes (univ. degree)	No	Yes
Rules and policies exist to evaluate auditor's work and skills	Yes	Yes	Internal of CB	Internal policies	Yes	Yes
Responsible entity exists to evaluate auditors	Yes, every 3 years	Yes	Yes, after planning or complaint	Yes, annual	Yes every 2.5 years	Yes, frequent
Certification body complies to ISO 65 requirements	Yes	No	Yes	No	Yes	Yes

Source: Majority of the data was obtained from the voluntary standard analysis and research of the ITC (International Trade Center) (http://search.standardsmap.org) and data for C.A.F.E. was extracted from the Conservation International website (http://www.conservation.org/campaigns/starbucks/Pages/CAFE_Practices_Results.aspx)

nal independent certifiers must be accredited by ISO and must conform to the specific requirements in the ISO 65. The use of external certification body ensures the independence of the certifiers from the governance mechanism of 4C. Similar to FLO or UTZ, the 4C also uses local or locally operating companies, but they do not use multi-stakeholder processes such as one followed by FLO. 4C auditors conduct surprise audits to ensure consistent adherence to standards. Similar to UTZ, the auditor needs to go through a training program to qualify for audits, but for a specific education background of the auditor is not necessary. 4C also employs rules and policies to evaluate the auditor's work and skills.

- *USDA Organic.* The USDA Organic audits are conducted by third-party certifiers or certifying agents. Only certifiers that have been accredited directly or authorized by the USDA are allowed to verify compliance to the NOP and certify products. The accreditation is valid for two and a half years and afterward the certifiers have to reapply for accreditation. Similar to other certifications, certifying agents have to comply with ISO 65 requirements. Certifiers that are authorized by USDA are certifiers accredited by foreign governments that have recognition agreement with USDA. These certifiers can be private, foreign, or state entities that are located around the world. USDA has accredited and authorized 84 certifiers, with 49 certifiers located in the United States and 35 certifiers in other countries. The certification process conducted by third-party certifier consists of five steps: (a) application by applicant, (b) desk audit by certifier, (c) field inspection by inspector,[19] (d) review of inspection report and documents, and (e) issuance of organic certificate by certifying agent/certifier issues organic certificate.[20] The USDA also does not require specific educational background, but inspectors need to participate in a training program to qualify for inspection. In addition, USDA also applies rules and policies to evaluate the inspector's work and skills.
- *C.A.F.E:* C.A.F.E. Practices relies on the SCS (Scientific Certification System) to administer and enforce their certification process. The SCS accredits third-party certifiers to act as certifying agents on behalf of the SCS. The SCS and the third-party certifiers need to comply with the requirements of ISO 65. The SCS also requires that the auditors employed by the third-party certifiers take training program and have specific educational background to qualify for audit. The SCS also frequently evaluates the auditor's work and skills.[21] The guidelines in C.A.F.E. Practices consist of 249 indicators. These indicators are used to evaluate the social and environmental performance of applicant (Semroc et al., 2012). C.A.F.E. Practices employs four different degrees of assessment results: noncompliant, verified, preferred, and strategic. Noncompliant is a condition when the applicant failed to meet the minimum requirements. Verified is assigned to applicant who met the minimum requirements and achieved a score of less than 60 %. Preferred is assigned if the applicant achieves a score between 60 and 80 % and strategic if over 80 % (Starbucks, 2013).

[19] The term "inspector" in the USDA Organic is comparable to the term "auditor" for fair trade.

[20] www.ams.usda.gov

[21] www.intracen.org

In general, the certifying agents use three strategies to assert the trustworthiness of their certification process. First, they ensure the independence of their certification process by eliminating conflict of interest between the certification agents and standard-setting bodies. Independency is in general assured by using independent (external) auditors and separating auditors from those that make final decisions about granting certifications. Second, the assessment process involves local knowledge and resources, especially during field audit. Most certification schemes also advocate connection with local knowledge and context by either involving local auditors or using multi-stakeholder process to develop audit indicators. Finally, the certification schemes use different strategies to emphasize the strength of their audits such as alternating between annual and surprise audits.

In sum, different certification schemes showcase different strictness in their assessment process to demonstrate the trustworthiness of their claims to the consumers. The existence or nonexistence of these assessment indicators, such as presented in Table 4.3, could signify different degrees of trust. For instance, full independence of certification body from the standard-setting body and full independence of the auditor from the certification body are argued to guarantee fairer audit results and limit possibility of collusion and manipulation (Deaton, 2004). Unfortunately, while these processes are known to the applicants, they are generally not known to the consumer. As a consequence, the ability of consumers to discern the trustworthiness of a certification and meaning of a label remains limited.

4.7 Concluding Remarks

Certifications and labels in the certified coffee market enable companies to provide its consumers non-price information, such as information about products' environmental and social sustainability. They are used to differentiate among companies based on their conduct and as a way for companies to limit competition or gain competitive advantage (Bartley, 2007). Endorsement by an external and independent organization is assumed to create trust among consumers who tend to disregard company's own claims regarding their ethical and sustainable conduct. However, as certifications and labels proliferated, consumers' trust in the validity of various seals slowly decreased. The decline in trust was caused mainly by the difficulty faced by consumers in verifying information behind the certifications and in the increasingly complex nature of the certification environment. In other words, the large number of certification and labeling schemes presents consumer with too many alternatives to choose from and obscures the meaning behind each individual label.

A close look at the six major coffee certification initiatives presented above reveals that in order to reassert their trustworthiness in the eyes of the general public, certifications emphasize transparency, legitimacy, and accountability of their practices. Majority of the certification and labeling schemes openly publicize their standards and principles to demonstrate the transparency of their governance process. They also use a number of approaches to assert the legitimacy of their practice, such as accreditation from reputable national or international organization. They also put emphasis on the democratic nature of their governance processes, such as strong collaboration with NGOs and producers during the standard-setting phase. The six certification and labeling schemes also demonstrate the accountability of their practice by emphasizing the independent nature of their certification agents, frequently evaluating their inspectors/auditors, and establishing formal mechanisms for complaints against the standard-setting body and certification agents.

Unfortunately, information about the steps certifications are taking to increase their trustworthiness is not readily apparent to the consumer. The information is unavailable either because consumers need to expand significant effort to research this information or some of this information might be proprietary. In addition, even if this information was available to the consumers, the magnitude of this information might deter consumers from using it as a basis for their purchasing decisions. We propose that one way to address these issues is to build an interoperable data platform that would enable private sector actors to share information and data through the use of agreed-upon semantic and ontology standards. Such platform can enable data owners and producers to make their information readily available by standardizing and simplifying the process for publishing information using semantic web-based technology. By making such information standardized and semantically interoperable, such platform would also enable social entrepreneurs to build decision assistance tools that are designed to empower a consumer to make a purchasing decision consistent with their values.

As argued in the previous chapter, considerable collaboration and trust building among public and private entities are crucial to push for private transparency to enable choice architecture of product information for the benefit of public. In addition, publishing the information also needs to take into account the three exceptions to information disclosure which are privacy, secrecy/confidentiality/proprietary, and national security (Stiglitz, 1998). One key problem with pushing private sector to share information is that the relevant data is often considered to be confidential, proprietary, and in some cases private data. In addition, considering the complexity of certification process, ensuring quality and security of the data and information for consumer use becomes crucial. The subsequent chapter reports on the creation of ontology-enabled interoperable data infrastructure based on semantic technologies to chare trust information of certification and labels, and Chap. 6 outlines the issues of privacy, secrecy, and security in information disclosure of commercial data and existing mechanism and strategies for negotiating these issues.

References

Albersmeier, F., Schulze, H., Jahn, G., & Spiller, A. (2009). The reliability of third-party certification in the food chain: From checklists to risk-oriented auditing. *Food Control, 20*(10), 927–935.

Ammenberg, J., Wik, G., & Hjelm, O. (2001). Auditing external environmental auditors—investigating how ISO 14001 is interpreted and applied in reality. *Eco-Management and Auditing, 8*(4), 183–192.

Bacon, C. (2005). Confronting the coffee crisis: Can fair trade, organic, and specialty coffees reduce small-scale farmer vulnerability in Northern Nicaragua? *World Development, 33*(3), 497–511. http://doi.org/10.1016/j.worlddev.2004.10.002.

Bartley, T. (2007). Institutional emergence in an era of globalization: The rise of transnational private regulation of labor and environmental conditions. *American Journal of Sociology, 113*(2), 297–351. http://doi.org/10.1086/518871.

Bartley, T. (2011). Mapping the hard/soft law terrain: Labor rights and environmental protection: Transnational governance as the layering of rules: Intersections of public and private standards. *Theoretical Inquiries in Law, 12*, 517–665.

Büthe, T. (2010a). Global private politics: A research agenda. *Business and Politics, 12*(3), 1–24. http://doi.org/10.2202/1469-3569.1345.

Büthe, T. (2010b). Private regulation in the global economy: A (p)review. *Business and Politics, 12*(3). http://doi.org/10.2202/1469-3569.1328.

Carrigan, M., & Attalla, A. (2001). The myth of the ethical consumer – Do ethics matter in purchase behaviour? *Journal of Consumer Marketing, 18*(7), 560–578. http://doi.org/10.1108/07363760110410263.

Carrington, M. J., Neville, B. A., & Whitwell, G. J. (2010). Why ethical consumers don't walk their talk: Towards a framework for understanding the gap between the ethical purchase intentions and actual buying behaviour of ethically minded consumers. *Journal of Business Ethics, 97*(1), 139–158.

Cashore, B. W., Auld, G., & Newsom, D. (2004). *Governing through markets: Forest certification and the emergence of non-state authority*. New Haven, CT: Yale University Press.

Castaldo, S., Perrini, F., Misani, N., & Tencati, A. (2009). The missing link between corporate social responsibility and consumer trust: The case of fair trade products. *Journal of Business Ethics, 84*(1), 1–15.

DeAngelo, L. E. (1981). Auditor size and audit quality. *Journal of Accounting and Economics, 3*(3), 183–199.

Deaton, B. J. (2004). A theoretical framework for examining the role of third-party certifiers. *Food Control, 15*(8), 615–619. http://doi.org/10.1016/j.foodcont.2003.09.007.

Dimitri, C., & Oberholtzer, L. (2009). *Marketing U.S. organic foods: Recent trends from farms to consumers* (No. 58). U.S. Department of Agriculture, Economic Research Service, United States.

Duff, A. (2004). *AUDITQUAL: Dimensions of audit quality*. Edinburgh, Scotland: Institute of Chartered Accountants of Scotland. Retrieved from http://icas.org.uk/res_duff_auditqual_report.pdf

Ecolabelindex. (n.d.). *Ecolabel Index|Who's deciding what's green?*. Retrieved August 12, 2013, from http://www.ecolabelindex.com/

Gibson, C. C. (1999). *Politicians and poachers: The political economy of wildlife policy in Africa* (transferred to digital print). Cambridge: Cambridge University Press.

Hatanaka, M., Bain, C., & Busch, L. (2005). Third-party certification in the global agrifood system. *Food Policy, 30*(3), 354 369. http://doi.org/10.1016/j.foodpol.2005.05.006.

Hirschman, A. (1970). *Exit, voice and loyalty: Responses to decline in firms, organizations, and States*. Cambridge, MA: Harvard University Press.

Howard, P., & Allen, P. (2010). Beyond organic and fair trade? An analysis of ecolabel preferences in the United States. *Rural Sociology, 75*(2), 244–269. http://doi.org/10.1111/j.1549-0831.2009.00009.x.

Jahn, G., Schramm, M., & Spiller, A. (2005). The reliability of certification: Quality labels as a consumer policy tool. *Journal of Consumer Policy, 28*(1), 53–73.

Janssen, M., & Hamm, U. (2011). Consumer perception of different organic certification schemes in five European countries. *Organic Agriculture, 1*(1), 1–13.

Kolk, A. (2010). Mainstreaming sustainable coffee. *Sustainable Development*. Retrieved from http://onlinelibrary.wiley.com/doi/10.1002/sd.507/full

Mayer, F., & Gereffi, G. (2010). Regulation and economic globalization: Prospects and limits of private governance. *Business and Politics, 12*(3). http://doi.org/10.2202/1469-3569.1325.

Muradian, R., & Pelupessy, W. (2005). Governing the coffee chain: The role of voluntary regulatory systems. *World Development, 33*(12), 2029–2044. http://doi.org/10.1016/j.worlddev.2005.06.007.

Panhuysen, S., & van Reenen, M. (2012). *Coffee barometer 2012*. Report, The Hague, Netherlands. Retrieved from http://www.teacoffeecocoa.org/tcc/Media/Files/Coffee-Barometer-2012

Pelsmacker, P. D., Janssens, W., Sterckx, E., & Mielants, C. (2005). Consumer preferences for the marketing of ethically labelled coffee. *International Marketing Review, 22*(5), 512–530. http://doi.org/10.1108/02651330510624363.

Pierrot, J., Giovannucci, D., & Kasterine, A. (2011). *Trends in the trade of certified coffees* (Technical Paper No. MAR-11-197.E, p. 17). Geneva: International Trade Centre.

Pivato, S., Misani, N., & Tencati, A. (2008). The impact of corporate social responsibility on consumer trust: The case of organic food. *Business Ethics: A European Review, 17*(1), 3–12. http://doi.org/10.1111/j.1467-8608.2008.00515.x.

Polonsky, M., Bhaskaran, S. and Cary, J. (2005), Exploring the opportunities for sustainable food labelling : a supply chain perspective, in ANZMAC 2005 : Proceedings of the Australia and New Zealand Marketing Association Conference 2005, ANZMAC, [Perth, W.A.], pp. 184–190.

RAN. (2012, April 11). The rainforest alliance demonstrates significant growth in 2011|Rainforest alliance. Retrieved July 29, 2013, from http://www.rainforest-alliance.org/newsroom/news/annual-growth-2011

Rao, H. (1994). The social construction of reputation: Certification contests, legitimation, and the survival of organizations in the American Automobile Industry: 1895–1912. *Strategic Management Journal, 15*(S1), 29–44. http://doi.org/10.1002/smj.4250150904.

Raynolds, L. T., Murray, D., & Heller, A. (2007). Regulating sustainability in the coffee sector: A comparative analysis of third-party environmental and social certification initiatives. *Agriculture and Human Values, 24*(2), 147–163.

Semroc, B., Baer, E., Sonenshine, J., & Weikel, M. C. (2012). *Assessment of the Starbucks coffee and farmer equity (C.A.F.E) practices program FY08 – FY10*. Starbucks & Conversation International, http://www.conservation.org/publications/Documents/2012_04_CAFE_Practices_Assessment_FY08-FY10_Final.pdf

Seyfang, G. (2005). Shopping for sustainability: Can sustainable consumption promote ecological citizenship? *Environmental Politics, 14*(2), 290–306. http://doi.org/10.1080/09644010500055209.

Starbucks. (2013). *C.A.F.E practices verifier and inspector operations manual – Starbucks coffee company* (Guideline No. V5.1). SCS Global Service, https://www.scsglobalservices.com/files/cafe_man_verinsops_v5_2_010614.pdf

Stiglitz, J. E. (1998). Distinguished lecture on economics in government: The private uses of public interests: Incentives and institutions. *The Journal of Economic Perspectives, 12*(2), 3–22. http://doi.org/10.2307/2646959.

Swaen, V., & Chumpitaz, R. C. (2008). Impact of corporate social responsibility on consumer trust. *Recherche et Applications En Marketing (English Edition), 23*(4), 7–34. http://doi.org/10.1177/205157070802300402.

Szmigin, I., Carrigan, M., & McEachern, M. G. (2009). The conscious consumer: Taking a flexible approach to ethical behaviour. *International Journal of Consumer Studies, 33*(2), 224–231. http://doi.org/10.1111/j.1470-6431.2009.00750.x.

Tanner, B. (2000). Independent assessment by third-party certification bodies. *Food Control, 11*(5), 415–417. http://doi.org/10.1016/S0956-7135(99)00055-9.

Taylor, P. L., Murray, D. L., & Raynolds, L. T. (2005). Keeping trade fair: Governance challenges in the fair trade coffee initiative. *Sustainable Development, 13*(3), 199–208. http://doi.org/10.1002/sd.278.

Weiss, J. A. (2002). Public information. In L. M. Salamon (Ed.), *The tools of government: A guide to the new governance* (pp. 217–254). Oxford, MA/New York: Oxford University Press.

Wells, J. T. (2002). Let them know someone's watching. *Journal of Accountancy, 193*(5), 106–110.

Chapter 5
Using Ontologies to Develop and Test a Certification and Inspection Data Infrastructure Building Block

Joanne S. Luciano, Djoko S. Sayogo, Weijia Ran, Nicolau Depaula, Holly Jarman, Luis F. Luna-Reyes, Giri K. Tayi, Jing Zhang, Jana Hrdinova, Deborah Lines Andersen, David F. Andersen, and Theresa A. Pardo

Abstract Global markets for information-intensive products contain sharp information asymmetries that lead to market inefficiencies resulting from consumer purchasing decisions that are based on incomplete information. Elimination or reduction of such information asymmetries has long been the goal of governments as well as various nongovernmental entities that recognize that addressing issues such as sustainable production, socially just labor practices, and reduction in energy needs and health expenditure is closely linked to consumers being fully aware of the economic, environmental, and social impacts of their purchasing decisions. This chapter reports on the creation of ontology-enabled interoperable data infrastructure based on semantic technologies that would enable information sharing in traditionally information-restricted markets. The main technical result is a proof-of-concept set of data standards built on semantic technology applications and the functionalities of formal ontology of certification and inspection processes. The current proof of concept focuses specifically on certified fair-trade coffee, and while its applicability is currently limited, it has the potential to become universally applicable to any certification

J.S. Luciano
Tetherless World Constellation, Rensselaer Polytechnic Institute, Troy, NY, USA
e-mail: jluciano@gmail.com

D.S. Sayogo
Department of Economics and Business, The University of Muhammadiyah Malang, Indonesia

Center for Technology in Government, University at Albany, Albany, NY, USA
e-mail: dsayogo@ctg.albany.edu

W. Ran (✉)
System Dynamics Society, University at Albany, 135 Western Ave., Albany, NY 12222, USA
e-mail: wran@albany.edu

N. Depaula • L.F. Luna-Reyes • D.L. Andersen
Department of Informatics, University at Albany, Albany, NY, USA
e-mail: ndepaula@albany.edu; lluna-reyes@albany.edu; dla@albany.edu

H. Jarman, L.F. Luna-Reyes (eds.), *Private Data and Public Value*,
Public Administration and Information Technology 26,
DOI 10.1007/978-3-319-27823-0_5

and inspection process for any product and service. In addition to producing a number of artifacts relevant to the expandability of the work, such as domain ontologies, the research indicates that while big data systems are necessary, they are not sufficient to create high levels of consumer trust. By testing the criteria using both hand-generated and automated queries, we are able to demonstrate that CIDIBB (Certification and Inspection Data Infrastructure Building Block) is not only able to test the trustworthiness of certification schemes but also that our ontology generates consistent results.

Keywords Ontology • Virtual certificates • Certificates • Semantic applications • Ontology validation • Certification ontology

5.1 Introduction

Economic theory tells us that a market clears when the upward sloping supply curve and the downward sloping demand curve cross—it is the basis for the bold assertion that free markets are the best way to distribute the factors of production to create a globally efficient production and distribution system. But hidden behind the theory of markets are assumptions about perfect information—both sellers and buyers must have access to the same perfect information about the state of the market. As we all know, these assumptions about information in free markets are often not true. When we choose (purchase) surgeon's services, we are often ignorant of how successful she has been in past surgeries. When we purchase a health insurance policy, we most likely do not know or understand what important features or services are not covered. When we buy a pair of running shoes, we do not know if they were

H. Jarman
Department of Health Management and Policy, School of Public Health, University of Michigan, Ann Arbor, MI, USA
e-mail: hjarman@umich.edu

G.K. Tayi
Department of Information Technology Management, University at Albany, Albany, NY, USA
e-mail: gtayi@albany.edu

J. Zhang
Graduate School of Management, Clark University, Worcester, MA, USA
e-mail: jizhang@clarku.edu

J. Hrdinova • T.A. Pardo
Center for Technology in Government, University at Albany, Albany, NY, USA
e-mail: hrdinova.1@osu.edu; tpardo@ctg.albany.edu

D.F. Andersen
Department of Public Administration & Policy, Department of Informatics, University at Albany, Albany, NY, USA
e-mail: david.andersen@albany.edu

manufactured in a sweat shop using child labor. When we purchase a pound of coffee, we do not know if it was grown in a way that exploited farm workers or damaged some distant ecology or even used unhealthy pesticides. Each of these market transactions points to the problems of information asymmetry that pervade many contemporary global marketplaces.

Current trends in consumer markets involve a growing number of ethical consumers who are increasingly paying attention to information about where, when, how, and by whom our food, consumer, and durable goods are produced (Bray, Johns, & Kilburn, 2011; Goleman, 2009; Watts & Wyner, 2011). For instance, organic market penetration for fresh produce has grown to 12 % in the United States since the adoption of USDA organic standards (Dimitri & Greene, 2002); fair-trade markets have grown 20 % in Europe and 40 % annually in the United States and the Pacific Rim (Kim, Lee, & Park, 2010). Yet information needed by ethical consumers during the buying process is still rarely available (Graham & Haarstad, 2011). Moreover, market premiums for organic, fair-trade, or environmentally friendly products offer an incentive to "greenwash" products by adding product labeling that promises low to no environmental impact with the sole aim of increasing profitability for the manufacturer or retailer and introducing additional sources for information asymmetries into the market. As a measure of the problem, a 2010 survey by TerraChoice conducted in 24 stores in the United States and Canada claimed that more than 95 % of the 5300 products being observed commit at least one instance of greenwashing (Makower, 2010; TerraChoice, 2010).

To reduce information asymmetry in this particular area, governments, NGOs, and private organizations have created a third-party certification and labeling industry. The numbers of third-party labeling initiatives have expanded rapidly since the 1990s (Albersmeier, Schulze, Jahn, & Spiller, 2009; Jahn, Schramm, & Spiller, 2005). The rapid proliferation of labeling obstructs the ability of consumers to observe the meaning behind labels, making their warranty of trusted information no longer adequate (Jarman et al., 2011).

This chapter describes the design and development of a semantic web-based prototype that could help correct information asymmetries in free markets. We are seeking to build an information infrastructure that can promote what we refer to as Full Information Product Pricing (FIPP) systems. FIPP systems are characterized by features that allow a surgeon an insurance company or a coffee producer to charge a premium (a FIPP price) for a product or service because consumers trust the information provided to them in regard to the key attributes of these products and services and these attributes align with consumers' values.

The rest of the chapter is organized in six sections following this introduction. Section two provides a vision of the ways in which CIDIBB can help in the creation of virtual certificates to promote FIPP systems. Section three describes previous research and some basic concepts about ontologies. Section four includes a brief description of the main components of CIDIBB as a set of three ontologies, CerTIN, FLO, and CiTruST. Section five presents an empirical evaluation of CIDIBB. In this section, we show ways in which CiTruST can be used to automatically classify certification systems in terms of their trustworthiness. Finally, section six includes a discussion, concluding remarks, and future work to fully develop CIDIBB.

5.2 Creating Virtual Certificates to Correct for Information Asymmetries in Markets

Our approach to producing FIPP systems involves the creation of a certification ecosystem. Certifying organizations will set explicit measurable standards for all types of products and services. Then third-party certifiers will send representatives to inspect facilities, processes, and outcomes to certify that they indeed meet the publicly available standards. Finally, a certificate will be attached to the product or service that gives consumers, or consumer advocates operating as agents of the consumers, all the trusted information that they need, i.e., perfect information without asymmetric bias.

Such systems already exist but only in partial form. Some organizations already produce certificates that are physically attached to products or services that we purchase. Examples of such physical certificates include Fair Trade and USDA Organic certificates that are printed on food products or the certificate issued by the local department of health of a clean and healthy kitchen hanging in many restaurants. Our scheme proposes that certificates, rather than being physically attached to products or services, become virtual certificates. Virtual certificates will be broadcasted on the internet and can be attached as an extended package of information to a unique product identifier such as a UPC product codes.

The challenge of making such virtual certificates a reality lies mainly in making the vast amounts of disparate data shareable and understandable across certification and inspection processes in a way that will be trusted by consumers. A key technical component that is necessary but missing is a combination of data standards and procedures that allow data to be shared seamlessly among the various and potential users of that data. We refer to this component as the "data infrastructure building block."[1] The Certification and Inspection Data Infrastructure Building Block (CIDIBB), whose creation is described in detail in this chapter, is a set of data standards in the form of a formal ontology of the certification and inspection process that would allow the creation of a data ecosystem for certification processes.

From a technical point of view, CIDIBB consists of four interlocking components that all operate in a semantic web environment. Three of the key components of CIDIBB consist of linked ontologies that model the domain of inspection and certification of consumer products with the certification and inspection of fair-trade coffee taken as a specific exemplar. The fourth component is a 28 question use case scenario that serves as a normative definition of what constitutes a trusted inspection virtual certificate (see Appendix).

At a high level, Fig. 5.1 shows the main components of the ecosystem created around CIDIBB. Such an ecosystem may include at least four classes of key

---

[1] The name "Data Infrastructure Building Block" derives from the National Science Foundation Data Infrastructure Building Block program which aims to "foster cross-community infrastructure development that solves common problems, while building blocks of data infrastructure that can support and provide data solutions to a broader range of scientific disciplines while reducing duplicative efforts." (http://www.nsf.gov/funding/pgm_summ.jsp?pims_id=504776)

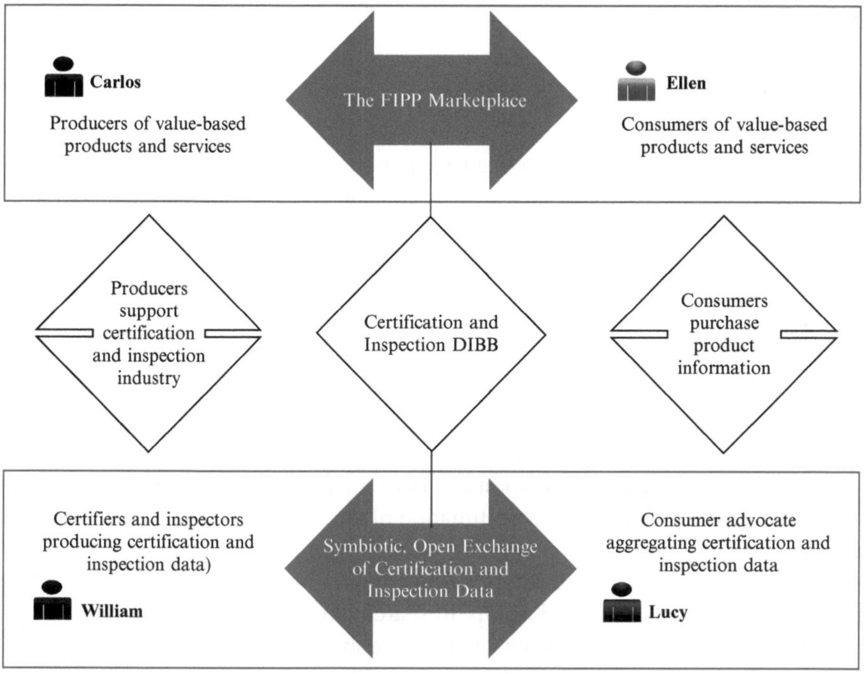

Fig. 5.1 Certification and Inspection Data Infrastructure Building Block (CIDIBB)

stakeholders that we represent by an idealized individual shown in each of the four corners of the figure.

Ellen, shown in the upper right-hand quadrant, represents consumers who will scan a product bar code to view its consumer rating and help them make a purchasing decision. Lucy, in the lower right-hand quadrant, represents a new consumer advocate industry that analyzes the full information package of consumer products and then sells that information to consumers such as Ellen. Lucy relies on William, a member of the inspection and certification industry, who uses CIDIBB to broadcast information about how, when, where, and by whom consumer products are created. The marketplace will drive what virtual certificates William is creating depending on what issues consumers are concerned with. For example, if consumers are concerned about the environmental impacts of the products they buy, then William's virtual certificates would focus on, for example, the carbon footprint created in producing and delivering the product to the final consumer Ellen. Carlos represents producers of value-based products and services. Carlos is cooperating with William to certify his production processes and to document unobservable attributes of his products because he understands that Ellen is willing to pay a price premium for products produced using methods that are congruent with her values. However, Carlos can continue to charge a price premium, and Lucy and William can stay in business only as long as Ellen continues to trust the information about virtual certificates that are being introduced into this newly formed Full Information Product Pricing (FIPP) ecosystem.

5.3 Previous Research

5.3.1 FIPP Systems and Trust

Our previous research has shown that trust plays a key role in all FIPP relationships (Luna-Reyes et al., 2013). In fact, trust is considered as an alternative governance mechanism in most collaborative relations (Powell, 1996). Higher trust levels lead to lowering of costs that result from the need to protect against opportunism (Shapiro, Sheppard, & Cheraskin, 1992). Moreover, the literature points out the importance of trust in these market transactions, particularly in the case of unobservable product attributes (Arora, 2006).

Researchers have identified several mechanisms for "trust production," which include calculative, relational, and institutional mechanisms (Rousseau, Sitkin, Burt, & Camerer, 1998). Institutional trust refers to the existence of an institutional framework that regulates the relationship between the main actors in the collaboration. Calculative trust refers to an estimation of the risks and payoffs intertwined in the interaction, and relational trust is associated with emotional bonds, shared values or objectives between the actors, or recognition of the trustworthiness of other participants in a repeated relationship. Research has found that institutional trust is particularly relevant for systems such as the proposed CIDIBB and that some features of information systems and information technologies contribute to building of institutional trust (Gefen et al., 2006; Luna-Reyes et al., 2013). Some of these features include peer feedback, online testimonials, affiliation links, guarantees, or system quality. The development of trust was central to designing the various components of CIDIBB, which are described in Sect. 5.4.

5.3.2 Ontologies and the Semantic Web

In the field of information and computer science, ontologies refer to explicit specifications of terms and their relationships within a domain of interest (Gruber, 1993). Such specifications provide a number of benefits, the most basic of which is enabling a computer to reason over the terms and properties of data (Uschold & Gruninger, 1996). Semantic web applications or services require that data be published in a format that makes use of these specifications as proposed in ontologies relevant to the domain of interest to which the data belongs (Berners-Lee, Hendler, & Lassila, 2001). Data published following such specifications may be called "linked data," and such data serve as building blocks for the semantic web (Berners-Lee, 2006). Creating data this way allows for more precise results from searches in the web and the automation of inferences over contents of the data (Bizer, Heath, & Berners-Lee, 2009). Specifically, using ontologies for the semantic web involves publishing data in the Resource Description Framework (RDF) file structure (W3C specification) in which subjects, predicates, and objects (or RDF triples) within components of the data are explicitly identified.

As semantic web technologies make use of specifications established in domain ontologies, they make it possible for data from different organizations and with different terminology—within a particular domain (e.g., certification and inspection processes) and using semantically equivalent concepts—to be integrated and classified in a structured way to improve searchability and the use of automated reasoning. For example, when a certification or inspection organization provides data where the "field inspector" is labeled as an "auditor," definitions in the ontology may indicate that these terms refer to the same concept, although they are labeled differently from one organization or one dataset to another. A software application can then use the ontology to determine that two attributes in two different datasets are equivalent. Applications can also use inference tools to make determinations about items and properties included in the dataset, such as "Is there an auditor?" or "Is the date of inspection before the date of certificate?"

The use of these tools and approaches makes it possible to adopt a framework that supports integration, data reuse, and automated reasoning. Therefore, these technologies are used in this research as the framework for our efforts to design, build, and test the concept of a Certification and Inspection Data Infrastructure Building Block (CIDIBB).

5.4 Key Technical Components of CIDIBB: Ontology-Based Data Standards and Evaluation System

CIDIBB is an abstract architecture for data storage, retrieval, and automated reasoning of certification and inspection data, based on semantic web principles. We developed three ontologies, CerTIN, FLO, and CiTruST (See Fig. 5.2). These three ontologies together form the fundamental base of the proposed Certification and Inspection Data Infrastructure Building Block (CIDIBB). CerTIN ontology defines the high-level abstraction of concepts, which we refer to as the global ontology. FLO ontology and CiTruST are called local ontologies. They inherit and expand high-level concepts defined in the global ontology. For more elaborate description about these ontologies and their development process, please refer to Sayogo et al. (Forthcoming).

CerTIN defines the most important and basic concepts of a certification system at a high level, meaning that CerTIN only provides the higher-level definition of a certification system that serves as an overarching architecture to connect multiple, more detailed ontologies for each certification and labeling scheme. The CerTIN ontology used standard definitions of class and property that are available from existing ontology literature. In addition, CerTIN has adopted classes and properties from three ontologies recommended by the W3C (World Wide Web Consortium).

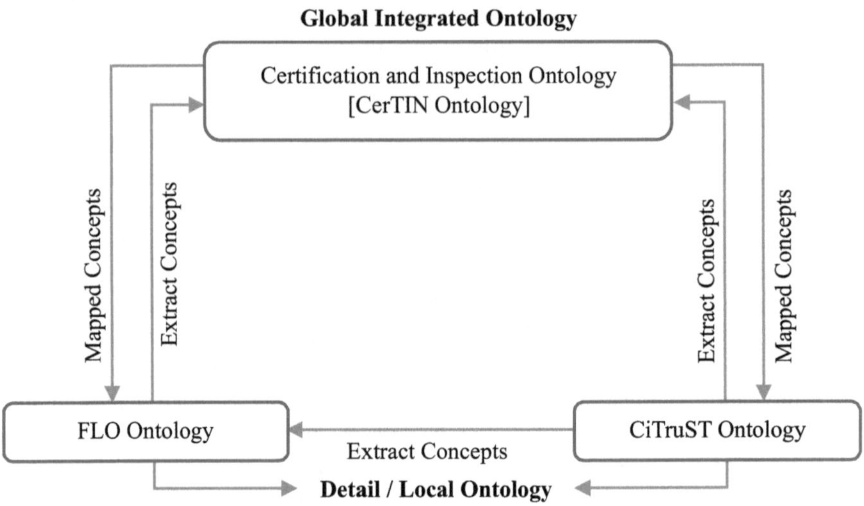

Fig. 5.2 The proposed ontologies and their relationships (Source: Sayogo et al. (Forthcoming))

These three include Dublin Core,[2] FoaF[3], and Good Relation[4] (Sayogo et al., Forthcoming).

CiTruST ontology was created to further demonstrate the scalability and expanded functionality of CerTIN ontology as an integrated global ontology. Thus, this ontology uses the classes and properties from CerTIN to define the quality of a certification process. CiTruST enables the determination of a "good" or "poor" certification process. We started with the basic structure of a certificate to find indicators for the quality of certification. Some components of the basic structure of a reliable certification process are accreditation body, certification body, standard setter, and monitoring process (Albersmeier et al., 2009; Deaton, 2004; Jahn et al., 2005; Tanner, 2000). The document analysis and interviews further indicated the importance of independence and monitoring processes that combine both document and field inspection as an indicator of reliable certification.

We thus posited that existence or nonexistence of particular components in the structure of certification indicates the degree of reliability of the certification scheme. The level of trustworthiness refers to the degree of certification trustworthiness derived from the conformance or nonconformance to the object of trustworthiness. The object of trustworthiness refers to the classes specified in CerTIN ontology. CiTruST ontology proposes four levels of certification process reliability, namely, level A to level D of certification trustworthiness. The assignment of the

[2] http://dublincore.org/2008/01/14/dcterms.rdf

[3] http://xmlns.com/foaf/spec/

[4] http://www.heppnetz.de/projects/goodrelations/

level depends on the existence of the criteria in the object of trustworthiness. The properties of level A of trustworthiness from CiTrusT ontology are shown below.

<'Level A of Trustworthiness'	*conformsTo*	***Some***	*Standard>*
<'Level A of Trustworthiness'	*certifiedBy*	***Only***	*'Third Party Certifier'>*
<'Level A of Trustworthiness'	*hasCertifyingOfficer*	***min 1***	*CertifyingOfficer>*
<'Level A of Trustworthiness'	*hasCompleted*	***Some***	*DocumentInspection>*
<'Level A of Trustworthiness'	*hasCompleted*	***Some***	*FieldInspection>*
<'Level A of Trustworthiness'	*hasEvaluationDecision*	***Some***	*CorrectiveMeasure>*
<'Level A of Trustworthiness'	*hasEvaluationDecision*	***Some***	*NonConformity>*
<'Level A of Trustworthiness'	*hasEvaluation Decision*	***Some***	*ObjectiveEvidence>*
<'Level A of Trustworthiness'	*hasStandardSetter*	***Exactly 1***	*StandardSetter>*
<'Level A of Trustworthiness'	*inspectedBy*	***min 1***	*Inspector>*
<'Level A of Trustworthiness'	*Authorize*	***min 1***	*Certificate>*

FLO ontology is an example of a local ontology that we created to further demonstrate how CerTIN ontology can be mapped to specific certification and inspection schemes. The ability of CerTIN to map into a local ontology such as FLO enables users (consumer advocates) to extract consistent and detailed information for assessing the trustworthiness of a certification scheme.

The most important elements of FLO ontology are the detailed classifications of compliance criteria into their properties. A compliance criterion is constructed with several restrictions, as defined in the FLO standard, by specific timeline, criteria types, measurement of the criteria, and organization applicability. These restrictions represent the properties of the criterion. Conformance to these properties affects the evaluation decision for certification, and it is also argued that conformance to these properties defines the level of trustworthiness of the certification schemes.

5.5 Empirical Testing of the Proof of Concept

After we created the basic building blocks of the CIDIBB, we devised a set of empirically based steps to test the resulting proof-of-concept system: (1) we generated a sample dataset drawn from the domain of Fair Trade certification of coffee grown in Mexico, (2) we used the ontology-based standards to publish this dataset in the form of an RDF triple store, (3) we generated SPARQL queries to determine if we could retrieve the answers to the 28 use case questions from the online published data (and if not, why not), (4) we summarized our results in terms of how many of the 28 use case questions could be answered with what level of assuredness and accuracy, and (5) we tested some limited reasoning capabilities to see if an inference-based system could distinguish between several datasets with known differences in quality and trustworthiness.

The testing process has been run against four datasets stored using the CIDIBB architecture: an Ideal Benchmark certification and inspection dataset and three certification schemes including FLO, Dave and Nic, and Nonviolent Dove. "Ideal Benchmark" certificate characterizes a hypothetical virtual certificate that could answer 100 % of the questions posed by the use case. FLO certificate dataset represents a high-quality virtual certificate. "Dave and Nic" and "Nonviolent Dove" are two synthetic certificates that represent degrees of greenwashed data. We created these two levels of greenwashed data by eliminating some key inspection data, not specifying criteria, or taking other shortcuts in the full certification and inspection process.

Answers to the 28 use case questions for each of the four datasets produced a unique distribution across the three classified levels (see Fig. 5.3). Differences in the level of difficulty required to retrieve the answers to these 28 questions can be used to assess the relative trustworthiness of various certification and inspection processes. Our test results clearly differentiate trustworthy virtual certificate datasets and datasets that yield less trustworthy virtual certificates.

Our results clearly distinguish between high-quality FLO data and data from the other virtual certificates that were missing answers to many of the detailed questions in the use case (See Fig. 5.3). We added an "Ideal Benchmark" certificate to characterize a hypothetical virtual certificate that could answer 100 % of the questions posed by the use case.

Tautologically, the Ideal Benchmark provides answers to all 28 questions in the use case, whereas the FLO certificate answers 19 of the questions; the lightly greenwashed certificate ("Dave and Nic") answers ten of the questions, and the heavily greenwashed certificate answers only seven of the detailed questions in the use case. Because the ontology contains an elaborated and semantically meaningful description of what a normatively defined good certification and inspection system should contain and because the use case questions do probe into some detail, greenwashed systems cannot "hide" the fact that their certificates are based on shortcuts and less than rigorous methods. Notice especially the sharp decline in questions that can be answered directly by SPARQL queries. By testing the criteria using both SPARQL and DL queries, we are able to demonstrate that not only is CIDIBB able to test the trustworthiness of certification schemes but also that our ontology generates consistent results.

Our empirical results show that the structure of CIDIBB provides a framework to evaluate the trustworthiness of different certification schemes. The CIDIBB architecture can support a system that integrates and exchanges massive amounts of

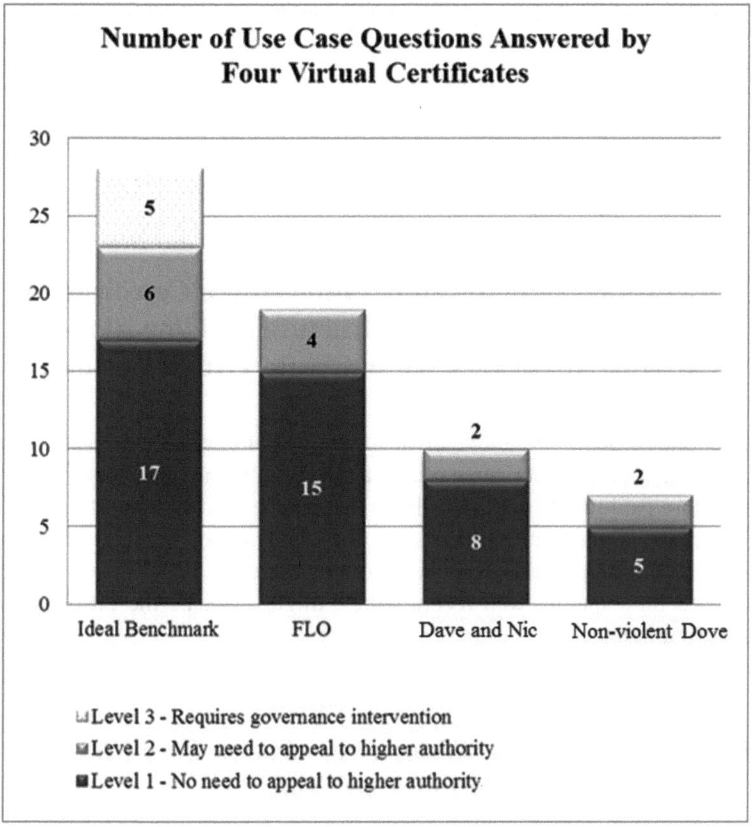

Fig. 5.3 The result of empirical testing of the certification schemes into the CIDIBB benchmark for trustworthiness

dispersed data about product certification and inspection schemes using semantically structured triple stores and allows consumer advocates, such as Lucy, to directly query such data for answers to the 28 use case questions and to use those answers to inform consumers, such as Ellen, about the trustworthiness of the certificates on the products and services she plans to purchase.

As discussed immediately above, a skilled human user of our CIDIBB can exhaustively query the existing data for multiple certification schemes, paying close attention to all 28 use case questions to arrive at the results presented in Fig. 5.3. Because of the use of semantic technologies, the manual process described above can be automated to classify a certification scheme as of four types (A through D) where an "A" classification is compatible with highly trusted data (again as defined by the 28 use case questions) and "D" classification is compatible with heavily greenwashed certification processes. Table 5.1 presents the results of our automatic classification of trustworthiness for the four datasets.

Table 5.1 The automated trustworthiness ranking of three certification schemes using the CiTruST ontology and reasoning

No	Certification scheme	Trustworthiness rating
1	FLO Labeling International (Flo-Cert)	A
2	Dave and Nic Certification	C
3	Nonviolent Dove Certification	D

5.6 Discussion: Vignettes Illustrating How a Certification and Inspection Data Infrastructure Building Block (CIDIBB) Might Be Used to Create Virtual Certificates

The components described above create a whole that is greater than the sum of its parts. These components constitute a "Certification and Inspection Data Infrastructure Building Block" (CIDIBB) that can be used to support the sharing of diverse datasets to meet multiple needs in the supply chain that feeds the consumer marketplace. We support this claim with four vignettes, illustrating how such a platform, based on semantic web technologies, might enhance the efficiency of the consumer marketplace.

5.6.1 Vignette #1: A Certifying Organization Uses CIDIBB to Create a New Virtual Certificate

William is the cofounder of CyberJustTrade (CJT), a start-up certification agency. As someone who is well trained in both sustainable marketing and economics, and who is knowledgeable about information systems, William understands the importance of differentiation strategy. Thus, he envisioned the creation of first-ever virtual sustainable certification scheme as the company's lever to compete against other much bigger certification agencies. His extensive observations and research on current certification schemes lead to an understanding of the lack of transparency in the current system. Based on this, he and his cofounder plan to enrich end consumers' purchasing experiences by providing a sustainable certification label that has three distinct functionalities: (a) enables instant traceability of certification information, (b) enables comparison against other certification schemes, and (c) provides increased transparency on certification information. William and his cofounder soon confronted three major challenges to their efforts: (a) the ownership of certification information is in the hands of the applicant and not the certification agency, (b) commercial privacy related to certification data for each firm in a supply chain, and (c) provision of instant traceability and comparability requires the availability of standardized data across supply chain firms and other certification schemes.

Upon discovering CIDIBB, William realizes it can help his certification agency in overcoming the abovementioned challenges. By requiring the applicant to store their data in CIDIBB compliant format, it will enable CJT to extract consistent and standardized information across the entire supply chain. Since CIDIBB is based on CerTIN global ontology, CJT could set up the level of abstraction of the data that needs to be extracted from the supply chain. By setting the level of abstraction to a higher level, CJT could evade the problem of commercial privacy. Since CIDIBB as a framework is equipped with well-established collaborative governance, the use of CIDIBB also solves the data ownership issue. The ability of CIDIBB to facilitate retrieval of consistent and standardized data supported with a trustworthiness ontology enables the CJT to easily compare the trustworthiness of different certification schemes against CJT certification and then to sell the comparison information.

5.6.2 Vignette #2: A Consumer Advocate Uses CIDIBB to Create a New Product or Service Rating System

Lucy is the CEO of a well-established product rating firm. Lucy's firm is an information aggregator that harvests information about sustainable consumer products and publishes proprietary product ratings (organized by UPC code). The firm has created a number of apps that allow consumers to access the product ratings, while they are shopping either in a physical store or online. Their business model is to sell a low cost-subscription to their service to individual consumers. One of the early entrants into their market niche was the GoodGuide product rating system.[5]

Figure 5.4 shows the architecture Lucy can use to build a CIDIBB-based product rating system. Consumer values are expressed as concerns and questions, which can be translated into machine-understandable queries. These queries are executed against standardized data and semantically enriched by CIDIBB ontology. For example, some consumers may be concerned if child labor was used during production processes (see Sect. 5.3). This concern can be translated into a machine-understandable query as presented below:

*If<NoEmployementOfChildrenUnderAgeOf15 hasEvaluationDecision**some**'EvaluationDecision'>and*

*<NoEmployementOfChildrenUnderAgeOf15 hasCriteriaType**value**"Core Criteria">and*

*<NoEmployementOfChildrenUnderAgeOf15 hasTimeline**value**"Initial Audit">and*

*<NoEmployementOfChildrenUnderAgeOf15 hasApplicability**value**"Members of Organization">and*

*<NoEmployementOfChildrenUnderAgeOf15 hasIndicator**value**"There are no Children under the age of 15 years employed">*

[5] GoodGuide is an actual product rating service that provides consumers with information about the health, environmental, and social performance of products and companies. http://www.goodguide.com

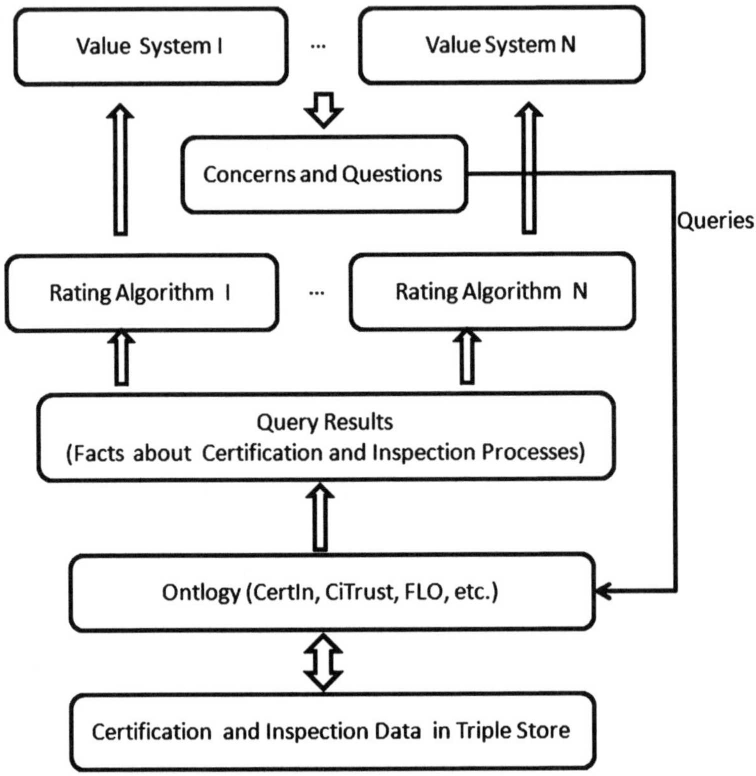

Fig. 5.4 CIDIBB-based product or service rating system

If the returned query result is true, then it means that no child labor was used in the production of the good. Query results are then fed into the rating algorithm. The output of the algorithm is one or more ratings that reflect the value of the good or service according to a particular value system.

5.6.3 Vignette #3: A Producer Featuring Sustainable Products or Services Creates Databases that Are CIDIBB Compliant

Carlos intends to reorient his mango producing farm in Central America to a farm that produces products that can be exported to premium consumer markets in the United States, Canada, and the EU.

Carlos recognizes that having multiple virtual certificates attached to his mangoes is key to the success of his export business. He wants his mangoes to be Fair

Trade certified, USDA organic certified, Shade Grown Certified, as well as being certified with a number of new virtual certificates that have emerged in the past several years (see Vignette #1 above). Carlos intends to manage all aspects of his business so that details of his operations, all of which are being inspected by multiple certification and inspection organizations, will all be as open as possible to anyone and everyone who is interested in buying his products. As Carlos contemplates setting up his business and its information systems, he recognizes that broadcasting the certification information of his mangoes to the internet using the CIDIBB will be key to his ability to charge a premium price.

In order to broadcast this information, Carlos has two options, (1) he can reconstruct his databases to make them CIDIBB compliant, or (2) he can retrieve the certification and inspection data from existing databases, further improve the retrieved data, and make them CIDIBB compliant. Choosing either option, Carlos will be facing some technical challenges.

5.6.4 *Vignette #4: TallMart Creates a Two-Sided Market Platform to Produce and Distribute Sustainable Products and Services*

The strategic planning unit of TallMart corporation, a major retailer in the United States and the EU, has realized that about 14 % of its base market consists of "green consumers," individuals who are willing to pay a premium for products that they can trust have been produced in conditions that are consistent with their values.[6] TallMart recognizes the potential of the CIDIBB to bring trusted information into the consumer marketplace as well as the commercial potential of creating a "two-sided" marketing platform wherein retail consumers pay a premium for products that can be sold with CIDIBB-certified virtual certificates, while at the same time producers of sustainable products are willing to pay a fee to have information about their products distributed on TallMart's platform using the CIDIBB standard, as shown in Fig. 5.5.

5.7 Concluding Remarks

Global markets for information-intensive products contain sharp information asymmetries that lead to market inefficiencies, resulting in consumer purchasing decisions that are based on incomplete information. Unintended side effects of these

[6] Walmart has been working with suppliers on various sustainability initiatives. In 2009 they introduced a sustainable measurement system that tracks the environmental impact of products. See especially: http://corporate.walmart.com/our-story/ or http://corporate.walmart.com/global-responsibility/

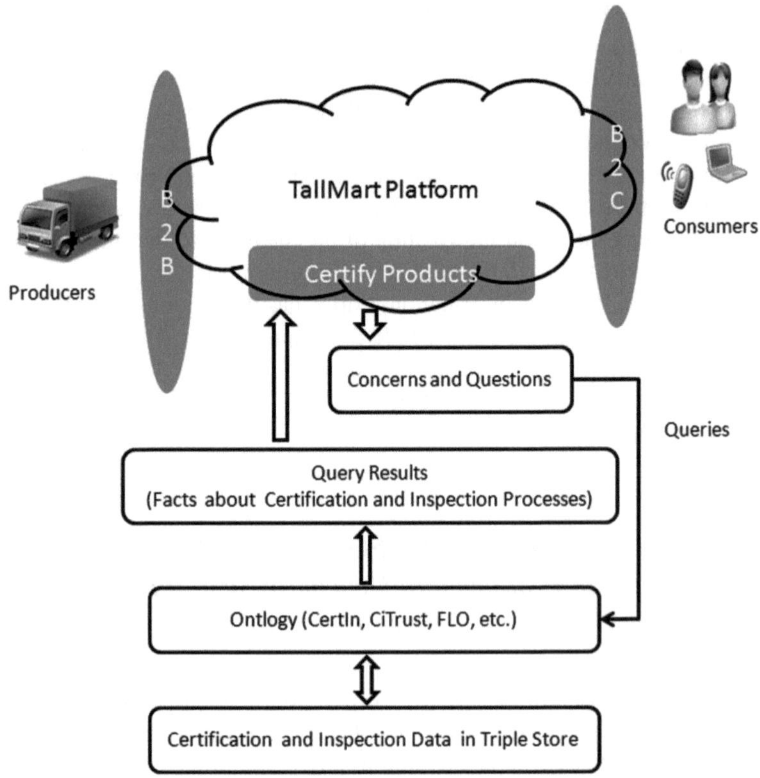

Fig. 5.5 TallMart platform

information asymmetries vary depending on the markets in question, ranging from negative externalities such as environmental degradation in the case of unsustainable production practices for agricultural products, loss of human capital in the case of exploitative labor practices, or unfavorable patient outcomes in the case of incomplete information about the quality of care provided in different healthcare settings. Elimination or reduction of such information asymmetries has long been the goal of governments as well as various nongovernmental entities that recognize that addressing issues such as sustainable production, socially just labor practices, and reduction in energy needs and health expenditure is closely linked to consumers who are fully aware of the economic, environmental, and social impacts of their purchasing decisions.

Our current research explored creation of ontology-enabled interoperable data infrastructure, based on the semantic web that would enable information sharing in traditionally information-restricted markets. Throughout the 3-year project, we explored the feasibility of tagging and broadcasting a diverse and dispersed set of data from product certification and inspection processes to allow for assessment of their accuracy and trustworthiness. The main technical result of our project is a proof-of-concept Certification and Inspection Data Infrastructure Building Block (CIDIBB), which is a set of data standards built on semantic web applications and the functionalities of a formal ontology of the certification and inspection process. While the current proof of concept focuses narrowly on certified fair-trade coffee and its functionality is limited, it has the potential for becoming universally applicable to any certification and inspection process for any product and service.

Achieving universal applicability of the CIDIBB, however, requires a series of steps aimed at refinement and broadening of our existing proof of concept and gradually increasing the scope of products and services. The first step is to further refine and test a full prototype in the original area of its focus, namely, certified fair-trade coffee. Such refinement and testing requires access to real-world certification and inspection data. The second step is the application of the refined CIDIBB to other certifications surrounding coffee, such as organic, to test the applicability of our Certification and Inspection Ontology (CerTIN) to other certification schemes. The continual focus on coffee takes advantage of our domain expertise and allows us to test CIDIBB's ability to address comparability of different certification schemes. If such buildup is successful, the next step toward testing for universal applicability is to incorporate other agricultural products that might require different inspection processes. Finally, the last step toward universal application is to use the existing CIDIBB for nonagricultural domains.

Making CIDIBB a reality requires integration of data and information that is under the ownership and stewardship of public and private entities. In this way, many nontechnical requirements also need to be met. While information quality and integrity have always been an issue of concern even in situations with a single information source, it will be an even more complex problem in the case of a platform that is designed to integrate information from multiple disparate sources. Thus, creating technical and process mechanisms to ensure information integrity and security is essential for the data to be trustworthy. Moreover, designing information policy that balances the need for supply chain transparency and ability of businesses to remain competitive is key. Establishing a governance structure is crucial for all large system development projects, but perhaps especially so for the development of platforms dealing with the complex determinants of sustainability such as CIDIBB. The key to this process is establishing a basis for "principled engagement"—a common understanding of the ways in which different stakeholders use central concepts and terms (Emerson, Nabatchi, & Balogh, 2012).

By making our proof-of-concept CIDIBB operational, we would provide, for the first time, a way for end users to reduce sharp information asymmetries in consumer markets through access to certification and inspection information in areas as widely dispersed as the performance of a healthcare provider, energy consumption patterns, or the safety of products we use each day in our daily routines.

References

Albersmeier, F., Schulze, H., Jahn, G., & Spiller, A. (2009). The reliability of third-party certification in the food chain: From checklists to risk-oriented auditing. *Food Control, 20*(10), 927–935.

Arora, R. (2006). Product positioning based on search, experience and credence attributes using conjoint analysis. *Journal of Product & Brand Management, 15*(4/5), 285–292. http://doi.org/10.1108/10610420610685695.

Berners-Lee, T. (2006, July 27). *Linked data.* Retrieved from Berners-Lee, T. (2006). *Linked data.* Retrieved from http://www.w3.org/DesignIssues/LinkedData.html

Berners-Lee, T., Hendler, J., & Lassila, O. (2001). The semantic web. *Scientific American, 284*(5), 34.

Bizer, C., Heath, T., & Berners-Lee, T. (2009). *Linked Data – The Story So Far: International Journal on Semantic Web and Information Systems, 5*(3), 1–22. http://doi.org/10.4018/jswis.2009081901.

Bray, J., Johns, N., & Kilburn, D. (2011). An Exploratory Study into the Factors Impeding Ethical Consumption. *Journal of Business Ethics, 98*(4), 597–608. http://doi.org/10.1007/s10551-010-0640-9.

Deaton, B. J. (2004). A theoretical framework for examining the role of third-party certifiers. *Food Control, 15*(8), 615–619. http://doi.org/10.1016/j.foodcont.2003.09.007.

Dimitri, C., & Greene, C. (2002). *Recent growth patterns in the U.S. organic food markets. Economic Research Service* (No. #AIB-777). Washington, DC 20036: US Department of Agriculture (USDA). Retrieved from http://www.ers.usda.gov/media/255720/aib777fm_1_.pdf

Emerson, K., Nabatchi, T., & Balogh, S. (2012). An integrative framework for collaborative governance. *Journal of Public Administration Research and Theory, 22*(1), 1–29. http://doi.org/10.1093/jopart/mur011.

Gefen, D., Pavlou, P. A., Benbastat, I., McKnight, H., Stewart, K., & Straub, D. W. (2006). ICIS panel summary should institutional trust matter in information systems research? *Communications of the AIS, 17*, 205–222.

Goleman, D. (2009). *Ecological intelligence: How knowing the hidden impacts of what we buy can change everything* (1st ed.). New York: Broadway Books.

Graham, M., & Haarstad, H. (2011). Transparency and development: Ethical consumption through web 2.0 and the internet of things. *Nformation Technologies & International Development, 7*(1), 1–18.

Gruber, T. R. (1993). A translation approach to portable ontology specifications. *Knowledge Acquisition, 5*(2), 199–220. http://doi.org/10.1006/knac.1993.1008.

Jahn, G., Schramm, M., & Spiller, A. (2005). The reliability of certification: Quality labels as a consumer policy tool. *Journal of Consumer Policy, 28*(1), 53–73.

Jarman, H., et al. (2011). *I-choose: Consumer choice, digital government, and sustainability in North America.* Presented at the APPAM Fall Research Conference. Washington, DC.

Kim, G.-S., Lee, G. Y., & Park, K. (2010). A cross-national investigation on how ethical consumers build loyalty toward fair trade brands. *Journal of Business Ethics, 96*(4), 589–611. http://doi.org/10.1007/s10551-010-0486-1.

Luna-Reyes, L. F., Zhang, J., Roy, R., Andersen, D. F., Whitmore, A., & Andersen, D. L. (2013). Information strategies to support full information product pricing: The role of trust. *Information Polity: The International Journal of Government & Democracy in the Information Age, 18*(1), 75–91.

Makower, J. (2010, November 1). *Is TerraChoice Greenwashing?* Retrieved from http://www.greenbiz.com/blog/2010/11/01/terrachoice-greenwashing

Powell, W. (1996). Trust-based forms of governance. In T. Tyler & R. M. Kramer (Eds.), *Trust in organizations: Frontiers of theory and research* (pp. 51–67). Thousand Oaks, CA: Sage Publications.

Rousseau, D. M., Sitkin, S. B., Burt, R. S., & Camerer, C. (1998). Not so different after all: A cross-discipline view of trust. *Academy of Management Review, 23*(3), 393–404. http://doi.org/10.5465/AMR.1998.926617.

Sayogo, D. S., et al. (Forthcoming). Ontological modeling of certification and inspection process to support smart disclosure of product information. *International Journal of Public Administration in the Digital Age.*

Shapiro, D. L., Sheppard, B. H., & Cheraskin, L. (1992). Business on a handshake. *Negotiation Journal, 8*(4), 365–377.

Tanner, B. (2000). Independent assessment by third-party certification bodies. *Food Control, 11*(5), 415–417. http://doi.org/10.1016/S0956-7135(99)00055-9.

TerraChoice. (2010). *Greenwashing Report 2010.* Retrieved from http://sinsofgreenwashing.org/findings/greenwashing-report-2010/index.html

Uschold, M., & Gruninger, M. (1996). Ontologies: Principles, methods and applications. *Knowledge Engineering Review, 11*(2), 93–136.

Watts, S., & Wyner, G. (2011). Designing and theorizing the adoption of mobile technology-mediated ethical consumption tools. *Information Technology & People, 24*(3), 257–280. http://doi.org/10.1108/09593841111158374.

Chapter 6
Privacy, Confidentiality, and Security Challenges for Interoperable Data Platforms in Supply Chains

Djoko S. Sayogo, Mahdi Mirdamadi Najafabadi, Giri K. Tayi, and Theresa A. Pardo

Abstract Privacy, confidentiality, and information security constitute basic requirements for the design and implementation of IT-enabled platforms for information sharing such as the I-Choose platform described in this book. In this chapter, we discuss privacy and security issues from an organizational perspective along three dimensions: ownership, access rights, and data quality. The challenge of protecting the confidentiality and privacy of data lies in developing effective and transparent security policies and protocols that govern access to and integrity of both proprietary and public information. Our findings highlight that these challenges stem from the complexity of the information chain and the heterogeneity of stakeholders and data sources in the sustainable coffee supply chain. As a result, addressing these issues will require not only technologically sophisticated solutions but also creation of governance structures and adoption of appropriate business practices. In this chapter we propose five management and policy solutions for mitigating the privacy,

D.S. Sayogo (✉)
University of Muhammadiyah at Malang, Jl. Raya Tlogomas no. 246, Malang, Indonesia

Center for Technology in Government, University at Albany,
187 Wolf Rd., Suite 301, Albany, NY 12205, USA
e-mail: dsayogo@ctg.albany.edu

M.M. Najafabadi
Department of Informatics, University at Albany, Albany, NY, USA
e-mail: mnajafabadi@albany.edu

G.K. Tayi
Department of Information Technology Management, University at Albany,
Albany, NY, USA
e-mail: gtayi@albany.edu

T.A. Pardo
Center for Technology in Government, University at Albany,
187 Wolf Rd., Suite 301, Albany, NY 12205, USA
e-mail: tpardo@ctg.albany.edu

H. Jarman, L.F. Luna-Reyes (eds.), *Private Data and Public Value*,
Public Administration and Information Technology 26,
DOI 10.1007/978-3-319-27823-0_6

confidentiality, and security challenges that confront successful implementation of I-Choose platform.

Keywords Privacy • Confidentiality • Information security • Data quality

6.1 Introduction

Smart disclosure represents a form of open data policy that has the objective of promoting innovations to help consumers make purchasing decisions that reflect consumer's social and environmental values (Thaler & Sunstein, 2008). In energy and utility marketplace, for example, an initiative called Green Button provides consumers with access to their energy consumption data, helping them to improve energy efficiency by providing them with "benchmarks and customized retrofit recommendations" (Thaler & Tucker, 2013). In the beauty products industry, "The Story of Stuff" campaign aims to make it a requirement for manufacturers to disclose information about the use of microbeads in their products. In recent years, the use of microbeads in personal hygiene products has been linked to the spread of microplastics throughout the marine environment (Thompson, 2004), which has been recognized as a serious, global environmental issue (Sutherland et al., 2010). There is now an international movement supported by 75 NGOs from 35 countries that uses a mobile application to classify products on the basis of their microbeads content.

Although much has been written about smart disclosure, the initiative is still in its formative stages from the consumer choice point of view. One of the barriers to an effective smart disclosure in the sustainable supply chain is the lack of integrated, quality information about the degree of sustainability for each of the interim products and raw materials. The main challenge is in capturing and amassing information across the supply chain while maintaining its timeliness, integrity, quality, and usefulness.

An interoperable data platform that leverages semantic web technology to connect different organizations in the supply chain, such as the one described in Chap. 5, is one possible solution to this challenge. Although there already exist private exchanges of information among participants in the supply chain supported by interorganizational information systems [1] (Choudhury, Hartzel, & Konsynski, 1998; Clemons & Row, 1993; Malhotra, Gosain, & Sawy, 2005; McLeod, Carpenter, & Clark, 2008), there is not yet an application that covers end-to-end information integration in a supply chain. Given the diversity of information and disparity of information sources in a product supply chain, such an interoperable data platform would require a complex sociotechnical system to integrate data under the stewardship of both public and private entities (Klein & Rai, 2009; Luna-Reyes et al., 2014; Steinfield, Markus, & Wigand, 2011).

Four elements of the information ecosystem contribute to the complexity of such supply chains: (1) information stored in diverse formats needs to be extracted and

[1] Examples of such interorganizational systems include efforts on Electronic Data Interchange (EDI), Electronic Markets, and Supply Chain Management Applications.

integrated from numerous and heterogeneous sources; (2) large volumes of data from heterogeneous sources and in disparate formats make analysis of such data very difficult; (3) some data might be available in a legacy form that cannot be used in automated analysis, and thus, conversion to machine-readable format would be required; and (4) some data, especially the data residing under custody of private entities, is mostly treated as nontransparent to the public—for reasons such as maintaining confidentiality and protecting data that is viewed by companies as proprietary. The same is true for data that is protected by individual privacy regulations.

Given this complexity, effective interoperable data architecture would at least require mechanisms that would guarantee legitimate access to all data needed, maintain and assess data integrity in order to generate trust among users, and create confidence in effective data protection measures for the data owner. This chapter outlines issues connected to securing authorized access rights, assessing data integrity, and the resulting challenges that might arise in creation of an interoperable data architecture for sustainable supply chain. Following the argument of Strong, Lee, and Wang (1997), this chapter uses the terms information and data based on their differences in the production process: data refers to "information at its early stages of processing," while information refers to "the product at a later stage."

This chapter only addresses information security challenges related to publishing and reusing data, leaving end-user security challenges and concerns (related to consumer use of smart applications using an interoperable architecture) outside the scope. We use the data collected as part of our project aimed to build an interoperable data platform for the certified sustainable coffee supply chain, as previously described in more detail in Chap. 5. This chapter is organized into seven sections including this foregoing introduction. Section 6.2 discusses information security challenges for an interoperable architecture based on a review of literature in the information system domain. Section 6.3 describes data complexities in the certified sustainable coffee supply chain. Section 6.4 brings up the issues related to securing legitimate access to the data. Section 6.5 focuses on data integrity and quality. Section 6.6 suggests five policy and management solutions to the challenges discussed in previous sections, and Sect. 6.7 concludes the discussion.

6.2 Challenges to an Interoperable Data Platform

There are numerous challenges in creating an interoperable data platform, ranging from unequal access to infrastructure and managing authorization, privacy, and confidentiality issues to making disparate data formats from various heterogeneous sources interoperable. In this chapter we focus on three specific challenges (adopted from the information system literature): securing legitimate access to data, governing access rights, and maintaining data quality and integrity (Rindfleisch, 1997).

Securing Legitimate Access A number of studies posit that the issue of securing access to information should be understood from a broader perspective and context (D'Aubeterre, Singh, & Iyer, 2008; Smith, Watson, Baker, & Pokorski,, 2007). Securing legitimate access to information in an interorganizational information system like the one proposed by our research requires a clear understanding of key security requirements such as confidentiality, integrity, and availability (Smith et al., 2007). Confidentiality refers to the protection of data and information from unauthorized disclosure. Integrity requires data to be protected from improper modification either on a server or in transit from one place to another in the system. Finally, availability refers to the continued accessibility to data and information services and exchanges. Meeting these requirements is already recognized as a key challenge of current information and data exchanges in supply chains (D'Aubeterre et al., 2008; Smith et al., 2007; Zhang & Li, 2006).

Governing Access Rights Maintaining access rights while securing legitimate access to the information is challenging because their goals can be seen as partially contradictory. Maintaining legitimate access might be affected if a large number of diverse entities are involved in the sustainable supply chain. As argued by Kagal, Finin, and Joshi (2003), large and heterogeneous entities are not expected to use similar data structures and data dictionaries (terminologies), thus making it difficult for their security protocols to communicate with each other. Similarly, when authentication processes receive large numbers of requests, they can become strained and thus become more prone to failure. Since authentication processes are necessary for any data-seeking activities, they can easily become a bottleneck for the whole system (D'Aubeterre et al., 2008). Therefore, the larger the number of entities involved, the greater the risk of failure to properly grant access to the requested data.

Depending upon the security architecture of the system, it can either lead to an inappropriate denial of access (because the authentication and authorizations mechanisms do not work properly and the system does not grant access to the information based on predefined access rights), or otherwise illegitimate access to information (since the system cannot determine the data seeker's access level using the authorization processes and the default access level has been set higher than the access level granted to the data seeker).

Maintaining Data Quality Data quality is known as the degree to which the data stored in the system is reflective of the truth in the real world. Some researchers argue that data quality needs to be understood within the context of "fitness for use" (Tayi & Ballou, 1998) or "utility" for a simpler term. Fitness for use stems from the relevance, interpretability, and ease of understanding of the information.

The literature has introduced several features associated with the quality of the data such as accuracy, timeliness, precision, reliability, currency, completeness, and relevance (Wang, 1998; Wang & Strong, 1996). Analyzing the importance of all these attributes from the user perspective, Wang and Strong (1996) grouped them in four distinctive categories relevant for our purposes. The first category involved data accuracy and included elements like the reputation of the source, its accuracy, and the believability of the data. The second category was related to the relevance of the data and involved categories such as the value added by the data to the user, its completeness, and its timeliness. Representation of the data was the third dimension and involved factors such as how easy it was to understand the data, as well as consistency and conciseness in its representation. The last category involved accessibility issues, which included knowledge in how to get access to the data in a cost-effective way.

An application of interoperable data platform along the entire supply chain necessitates integration of a large number of independent but semantically related data sources (Luna-Reyes et al., 2014). To better understand the integration complexities, we use the case of the certified sustainable coffee supply chain to discuss the challenges regarding securing legitimate access, governing access rights, and maintaining data quality.

6.3 The Information Chains in Certified Coffee Domain

As we have already described in Chaps. 2 and 3, the information needed to sustain transactions in a certified sustainable coffee supply chain involves multiple stakeholders using their own information systems that record and manage data in different formats. In general, there are two information chains in the certified coffee supply chain: sustainable trading information and the certification information. Sustainable trading information refers to the information regarding the trading of sustainable products. This information is contained in the trading documents such as invoices, bills of lading, farmer's contracts, and financial contracts. The information content of each document might differ for each firm. Certification information refers to the information collected by a certification body during the process of granting sustainable certificate to the companies in the supply chain.

The two chains are interrelated and provide a complementary picture of information flows in a certified coffee supply chain. For instance, certification ID (one record type among the certification data) is required in each document used in sustainable trading. Sustainable trading information is then compiled into quarterly reports submitted to the certification body to ensure compliance control over traded certified goods.[2] Thus, any data platform that aims to make smart disclosure for sustainably produced coffee needs to capture data from both information chains in order to be capable of building a trustworthy result.

The complexity of the information chains in the certified coffee supply chain calls for a more robust identification of the critical components of an interoperable data architecture. As discussed earlier, an interoperable data platform needs to capture and amass data that is aggregated and held by both public and private entities. As argued in other chapters of this book, triangulating the data collected from public entities and the data from private entities along the supply chain is needed not only to ensure comprehensiveness of the data but also to verify its reliability. This verification is needed to enable smart disclosure tools make comprehensive analyses and suggest trustworthy recommendations to end users.

Adapting from the White House's Memorandum on Informing Consumers through Smart Disclosure (Sunstein, 2012), we posit three different types of infor-

---

[2] Detailed description of data collection process can be found in documentations published by third-party certification bodies. For example, for FLO, refer to http://www.flocert.net/fairtrade-services/fairtrade-certification/how-it-works

mation that might be critical for an interoperable data architecture for certified sustainable supply chains:

Product information such as full pricing information, geographic availability, product origin, compliance standards, type of labels, features and attributes, nutritious contents, ingredients, and production process

Supplier/trader information such as types of certifications, location, in-trade associations, complaints, citations, financial situations, and producer's information

Third-party information including subjective ratings of the product, certification process, certification results, auditors, etc.

The complexity of the information chain and also the conflicting interests of different actors (e.g., openness versus privacy and confidentiality) makes capturing critical data components from the sustainable supply chain a challenge. In the following sections, we describe the challenges of securing access to data, managing access rights, and maintaining data quality and integrity in greater detail. We specifically focus on such challenges from the perspective of producers, roasters, and third-party certification bodies.

6.4 Securing Legitimate Access

Securing legitimate access to information for all actors involved in the certified sustainable coffee supply chain is difficult partially due to the competing nature of openness, commercial privacy and confidentiality, information availability and integrity, and the corresponding differences among actors' interests in their information disclosure policies. Some information about the product such as nutritious contents are now publicly disclosed by vendors due to government mandates, while withholding information such as supplier's location(s), production process, or the origin of raw materials are still justified by arguments of trade secrets and protection of firm's competitive advantage.

The data used in this section was collected through in-depth interviews with various stakeholders of the certified coffee supply chain—producers, exporters, importers, roasters, third-party certification bodies, and also some nongovernmental organizations. The interviews were conducted in two waves: from May until June 2012 and from November 2012 until May 2013. Additional secondary data was collected from openly available documentation from seven major coffee initiatives.[3] For a detailed description of all methods used in the book, refer to the methodological appendix.

[3] The seven major coffee certification initiatives: FLO, UTZ Good Inside, 4C, RAN, Organic, C.A.F.E Practices, and Nespresso AAA (Panhuysen & van Reenen, 2012).

6.4.1 Data Ownership

Information collection in the sustainable certified coffee supply chain is conducted by two different types of entities. Information owner, who is often also the producer of the information, generates and provides input data to the certification process and sustainable trading. Information owners in sustainable certified coffee supply chain are coffee producers (e.g., small farmers and cooperatives) and coffee traders (e.g., roasters or importers). The second group that collects information are information custodians/stewards, who receive, record, and maintain the information from the information owners. This role is generally fulfilled by the various certification bodies in the sustainable certified coffee trade such as certification authorities like FLO, UTZ, or the USDA, as well as certifiers who manage the inspection process (e.g., CertiMex, Bio Latina, etc.).

Although information custodians maintain the data in their systems, it is the information owner whose decision matters when it comes to making data available for a third-party agent. In other words, extracting data in the custody of information stewards requires consent from the information owner. Some of this information is already being shared in proprietary information systems such as the Ecert, which is the FLO platform to collaborate during the certification process. The existence of platforms such as Ecert facilitates the integration of the platform described in Chap. 5. However, confidentiality, integrity, and availability need to be warrantied for all participants.

6.4.2 Conflicting Disclosure Policies

As mentioned before, actors in the certified coffee supply chain have different information disclosure policies. For example, our interview with the director of Fair for Life certification body indicates that the disclosure policy of the certification applicant—the data owner—very often conflicts with the disclosure policy of the certification body and of course applicants take into consideration their own and unique data disclosure policies for publishing their certification data. The applicants are especially concerned with the impact of such data disclosure on their brand image and reputation. Conflicting disclosure policies pose challenges to consistent extraction and integration of data in the smart disclosure process. In its simplest form, a data owner might allow disclosure of a particular piece of information, whereas disclosure of the same piece of information is not allowed under policies imposed by another data owner. These discrepancies might hinder the ability and performance of the system to map and trace information from the entire supply chain. Again, extended conversations among actors are needed to clarify the types of data that can be shared among members of the supply chain and types of data that should be shared or opened to the general public.

6.4.3 *Assessing Data Integrity*

Maintaining the integrity of the data is crucial for any data platform since—as discussed earlier—there are uncertainties stemming from the sheer number and variety of data sources and agents. A large number of agents would be involved or affected by an interoperable data platform in the semantic web technologies (O'Hara, Alani, Kalfoglou, & Shadbolt, 2004). Consequently, the system should acquire access to numerous heterogeneous data sources for verification purposes, by triangulating and comparing instances of data from different sources (Huynh, Jennings, & Shadbolt, 2004). Once data sources are discovered and their integrity is approved, it will have to ascertain the relevance of the information.

Understanding the provenance of the information is necessary in every domain and more so in a supply chain certification and inspection domain due to the dependency of the pipelined data (which virtually moves along the entities in the supply chain) on time and context (D'Aubeterre et al., 2008). The inherent subjectivity in this domain could make the assessment of the trustworthiness of information and the legitimacy of the information producers more complicated. There might be situations in which one agent is assessed as trusted on a particular information but another agent is not. For instance, a certification generated by a nongovernmental organization (NGO) that does not have a conflict of interest might be viewed as more reliable than a certification generated by commercial brands such as Starbucks or Nestle.

6.5 Governing Access Rights

An important challenge for implementing an interoperable data platform lies in the ability to produce models of integrity assurance that recognize the context-sensitive information but at the same time recognize different access levels. Consequently, information policies capable of governing trust should be developed. Jarman et al. (2011) argue for the creation of data commons (see also Chap. 8 in this volume), a roundtable type of governance model involving stakeholders in supply chains, to generate policies to support the implementation of interoperable data platforms. This governance structure would need to involve a wide range of stakeholders including government regulators, industry associations, consumers and consumer advocates, producers, and others. The policies could govern the minimum level of data integrity that each agent in the supply chain must comply with to ensure trustworthiness of information. Additionally, the policies could specify the requirements for authentication and foolproof authorization of data seekers as well as data providers to suffice this aim.

6.5.1 Repercussions of Information Sharing

A data commons like the one briefly described in previous paragraphs implies the existence of a system with several levels of access. FLO Ecert information system is in a sense an example of a data commons. Ecert facilitates the workflow of the certification system, facilitating information exchange among certifiers and data owners. Most of this information, however, remains private. Sharing this information (or part of it) with the public or with other entities depends mostly on the perception of value creation. Private entities regard their information as strategic assets from which they can extract significant business value, as certain information is closely related to core competencies that lead to firm's competitive advantage. In this view, disclosing this information can benefit firm's rivals and potentially endangers the organization's market share. Therefore, information disclosure decision—i.e., granting access to information to a third party—is associated with the expectations of returns (Stiglitz, 2000). Interviews with mission-driven companies in the certified coffee industry point to concerns of economic repercussions in opening information. The interviewees indicated willingness to open their information only if doing so can add value to the organization and would not erode their competitive advantage as sources or enablers of revenue.

Opening other types of information about the firm and the product might also have impact on the market share. For example, disclosing some details of certification audit results might compromise the reputation of the applicant. Our interviewee from the Fair for Life certification body stated that companies are extremely cautious about what details of their certification audit results will be published, for fear of damage to their reputation and public image.

Addressing concerns related to economic repercussions of disclosing information might require information policies that offer an effective governance structure for not only promoting disclosure as a mechanism to deliver public value but also to have solutions for the potential challenges stemming out of concerns over integrity and confidentiality. Our interview findings indicate that when such policies are in place, disclosure is more welcomed since it creates value. For instance, in the case of mission-driven companies, opening their data to support smart disclosure is desirable if it allows publicizing their exceptional conduct of economic, social, and environmental sustainability. This is because although each organization assigns different value upon information from which they draw their competitive advantage, they are all seeking ways to promote their public image, while protecting their core competencies and continuing to generate revenue by utilizing them.

6.5.2 Challenges to Information Access

As we described early in this chapter, clear policies on access to shared data resources are a key requirement for a platform such as I-Choose. Information shared needs to be protected from potential unauthorized access as well as inference attacks.

Unauthorized access refers to a situation where a piece of information is accessed by a user who should not have access to it. In many instances, the same user might be authorized to have access to a subset of the same piece of information legitimately. Therefore, it is important not only to keep unauthorized people behind the authentication wall but also to have an effective access control mechanism in place to guarantee that it only grants access to the requesters as per their legitimate access rights, based on a preexisting agreement with the information owners or otherwise a strong information disclosure protocol. In some cases, an attacker might try to leverage its access rights by fooling the governance structure, and hence intrusion detection mechanisms can also be a part of the system to allow detecting even unknown malicious efforts which follow a known pattern. The system should encrypt the data at all communication and storage points, to minimize risks of unauthorized disclosure by eavesdropping or intercepting the connection to the data sources.

Inference attack refers to the ability of a user to deduce information, which they are not authorized to access. This is done by using the information that they have legitimate access to and using it to reproduce a higher level of information which was not the intention of the governance structure. There are two factors determining the amount of risk by inference attacks: the number of organizations involved in the certified coffee supply chain and the availability of large volumes of information (for legitimate users). For example, the certification body might remove all information but the certification ID from audit results of producers to maintain anonymity. However, by connecting certification ID with supplier information from roaster, an attacker might be able to identify the actual producer. Given the nature of the data collected, it is not always an easy task to determine the possibility of such attacks, and it is even harder to design and enforce governing protocols and policies to ensure such attacks are not feasible. However, there are known techniques in computer security inference control profession that are designed to ensure that one cannot infer new pieces of information from a certain number of interconnected datasets that she/he has legitimate access to but none can completely guarantee that such attacks are impossible in rare circumstances.

In this model, new governance mechanisms are needed in order to change the current mindsets of information owners and information disclosure policies in the certification context. According to the director of Fair for Life certification body,[4] disclosure of certification information is only on a voluntary basis under the current system, leaving the data owners with the option not to disclose their information (or at least certain parts of their information) to the public. Nonetheless, the new governance model should promote information policies that guarantee access to such information for the public or legitimate information seekers, while maintaining information owner's integrity and confidentiality.

In addition, the model should also include an authentication and authorization process to govern the appropriate access level. Granting the appropriate access level to data consists of two distinct processes: authentication process, which determines the identity of the data seeker, and authorization process, which grants an appropriate

[4] http://www.fairforlife.org

access level to the data seeker based on some predefined policies or otherwise stored data. The latter defines who can access what type of information (Rindfleisch, 1997). Under the current certification system, data owners can set different access rights and permission levels for different actors and can employ various solutions for authenticating data requesters. One such solution involves a formal agreement between data owners, potential entrepreneurs, and data custodians. In this model, the data owner provides the data custodian with a list of actors who are authorized to grant a given level of access to the data. Another solution would be to permit for instantaneous authorization, where the system authenticates the data requester, making the decision to allow or disallow access to the data right on the spot, based on a predefined authorization protocol.

The resulting variety of protocols for accessing data would impede the ability of the system users to access data from various data owners. To illustrate, consider the 444 ecolabels worldwide as identified by EcoLabelIndex.[5] If each of these certification bodies hosted information from just 100 coffee producers, there would be 44,400 data owners whose disparate authentication and authorization processes would have to accommodate to the interoperable data platform. Thus, the technical solution must include an efficient operation toward such diverse and heterogeneous system.

6.6 Maintaining Data Quality

While information quality and integrity has always been an issue of concern even in situations with a single source of information, it will be an even greater issue in a platform designed to integrate data in multiple formats from heterogeneous sources. In this section we discuss major challenges to maintaining information quality and integrity in this context.

6.6.1 Information Accuracy Challenges

In the certification process, two major factors can influence information accuracy: human judgment bias and actual data fabrication at the data source. Human judgment bias affects information accuracy as certification decisions are made based on field reports from certification auditors. Even though certification auditors undergo vigorous training, the element of human subjectivity cannot be fully eliminated given the interpretability of some of the certification criteria. There are at least two points where human judgment impacts the certification decision—field audit and evaluation interpretation of the audit results. For example, each of the Fairtrade Labeling Organization (FLO[6]) measurement criteria is based on five levels of com-

[5] http://www.ecolabelindex.com

[6] http://www.fairtrade.net

pliance; to acquire the certification, the applicant must meet at least the third level. However, the indicators for each compliance level are not always prescriptive. For instance, the third-level indicators for criteria 3.2.22 of FLO—training members on appropriate use of fertilizers—are (a) at least 50 % members have been trained and (b) content of training was sufficient. In this case, the auditor will have to use her judgment to measure the sufficiency of the training content. Such subjective judgment can distort the fact and might result in less objective and lower-quality information recorded (Strong et al., 1997). The subjectivity of these reports will in fact have an effect on the accuracy of information.

In addition to the human bias, the factual value of information can be distorted if the original data is wrong or is intentionally fabricated. Based on our interview with cooperatives, information fabrication can happen in the certified coffee supply chain due to financial interests, and the remoteness of certification applicant's location makes it more challenging to cope with. Our findings suggest that small farmers often sell their coffee to local intermediaries who in turn sell it to the cooperatives. According to the interviewees, local intermediaries often fabricate information about the origin or the certification status of the products in order to fulfill orders more quickly. This is mostly due to the exclusion of local intermediaries from disclosing the premium accrued from the certified coffee. As a consequence, although the quantitative pieces of data might remain correct, it usually does not represent the actual sales of the certified coffee.

The threat to information accuracy caused by human bias and data fabrication in the supply chain requires governance-based process controls. There are three possible approaches to creating such controls:

1. The certification body assures the competence of auditors by developing policies for auditor's training and requirements for specific educational background and experience levels.[7]
2. The certification body enforces the implementation of some better internal controls in the audit process at the applicant's level.
3. An expert system monitors and analyzes the audit results against actual data and utilizes historic patterns and heuristic algorithms, in an effort to minimize human bias.

6.6.2 Information Provenance Complexities

Provenance is fundamental to data quality as it relates to the relevance of the information in the context of our application (Buneman & Davidson, 2010). As argued by Buneman (2013), the process of making a platform, which is based on semantic

[7] Currently, only three out of seven major coffee certifications require specific educational background for their auditor. See Panhuysen and van Reenen (2012) or the International Trade Center (http://legacy.intracen.org/marketanalysis/Default.aspx).

web technologies, "provenance aware," is a nontrivial task. Understanding information provenance is important to ensure that the recommendations provided by smart disclosure tools using interoperable data platforms are relevant to the consumer and based on valid information. This is especially important in the sustainable coffee certification domain due to time-relevance and context-relevance nature of the information. Since the meaning of the information is influenced by time and/or context, ignoring these dimensions may distort the analysis and consequential recommendations. Here are two examples to illustrate the importance of the two dimensions.

First, standards and certification criteria are dynamic and therefore can be changed frequently. Thus, if the data platform does not consider provenance in terms of the versions of the standard and/or certification criteria, the recommendation provided by the decision support tools that draw information from such platform can become misleading after a certain amount of time. Our interviews with the Fair for Life certification body indicated that the Fair for Life decided not to publish certification results for more than 2 years back, as the frequent changes in standards make certification results incomparable to older records.

Second, in situations where organizations are switching from one certification to another, not knowing the context of the certification ecosystem can result in distortion of certification and decertification information provided on the certification website/database. For instance, Fairtrade Canada provides the listing of the registered and de-registered firms on their website. A firm de-registered by Fairtrade Canada at the end of 2012 might gain certification in Fairtrade USA the same year. Without knowing the provenance in term of context, the platform might equate Fairtrade Canada and Fairtrade USA as the same certification under Fairtrade. This misses the fact that Fairtrade USA is a totally different certification body whose certification criteria differ from those of Fairtrade Canada and abide by the criteria of the FLO.

6.6.3 *Data Representation and Reuse*

Guaranteeing safe and fair secondhand use of data after it is disclosed to a legitimate party or published openly is a major concern for both data owners and data custodians. Some certification and inspection data in certified coffee domain might fall under commercial privacy categories, such as trade secrets, that can directly affect the company's competitive advantage. Intrusion into the data repositories can be detrimental for the data owner since it may lead to destruction, unauthorized manipulation, or illegitimate disclosure of business information that could endanger the sources of revenue of the company and hence its business continuity. Alteration of the confidential information could impair the integrity of the certification and inspection information and may lead to mistrust in the company's image. Unauthorized manipulation of the data can be damaging for the data owners (i.e., coffee producers) and will increase data seekers' liabilities as well. Repudiation constitutes another potential problem that will have an impact on the trust in the data. For instance, a data owner might allow access to parts of its certification data but after some feedback from the market or after it was used as the basis of an assessment by a certification body, decide to deny (or repudiate) the original data or its consequential information in further stages of the supply chain.

As argued by Zimmerman (2007), appropriate metadata to enable tracing the provenance should be complemented with regulations and policies to ensure proper and accountable reuse of data. Therefore, corresponding legal and policy frameworks that ensure that data will not be poached and misused systematically if reused—as evidenced in the scientific research domain (Sayogo & Pardo, 2013)—should also be developed.

6.7 Mitigating Security Risks: Management and Policy Implications

Ensuring information security of an interoperable data platform should encompass not only technical elements to ensure access control and integrity but also integration of policies, business rules, management practices, and actors' culture. This section outlines the policies and managerial aspects to govern the information issues outlined above from the perspectives of data owners and the data seekers.

6.7.1 Integrating an Information Security Policies and Culture

Ensuring information security is not a solely technical issue (Lim, Chang, Maynard, & Ahmad, 2009). As we described in Chap. 2, sustainable certified coffee supply chains involve not only small or medium producers, cooperatives, and roasters but also large intermediaries and well-resourced enterprises such as Starbucks or Nestle. Each of these participants might have different information security and privacy policies in place or none at all. An assumption suggesting that most organizations have already developed information security and privacy policies to protect their information assets exposed security risks might not apply generally (Saint-Germain, 2005). Efforts should be made to ensure that participants in sustainable certified supply chains are all aware of the possibility of information loss and security breaches farther in the supply chain and the potential consequences.

As previous research shows, majority of information security breaches are caused by poor security behavior (Da Veiga & Eloff, 2010; Leach, 2003) such as noncompliance with information security policies (Pahnila, Siponen, & Mahmood, 2007). In an interoperable data platform where data sources and data seekers are heterogeneous, lapses of information security behavior by employees can be damaging because the magnitude of breach can be escalated. In this way, managers need to recognize the importance of fostering an information security culture by integrating information security and privacy policies into everyday business processes and procedures. It is crucial to change the mindset of the employees who might still believe that information security is the sole responsibility of a handful of technology products and the IT personnel. In fact, employees themselves are the component of information security systems where there is no known defense against risky actions such as downloading an attachment or following a link that may expose sensitive data to security vulnerabilities and threats.

Thus, integrating information security and privacy awareness along with information security best practices is necessary for companies in various stages of the supply chain. This, of course, will also require appropriate training of the actors involved in different stages of the coffee production supply chain and also the certification bodies. This requires an alignment of strategic choices and direction, business processes, and information security policies and practices.

6.7.2 Balancing Between Confidentiality, Commercial Privacy, and the Benefit of Openness

Protection of information is crucial to the continuity of a business organization. There are many cases where information breach resulted in significant business loss for an organization. Cavusoglu, Mishra, and Raghunathan (2004) assessed an average loss of 2.1 % of a company's market value within two days of publicly announcing a security breach.

On the other hand, opening data is valued by consumers. A thorough review by Healy and Palepu (2001) sheds light on the fact that numerous studies in accounting, management, and economics affirm the benefits of publicly disclosing information. Moreover, the basic assumption of the smart disclosure policy is in fact the creation of economic value for participants in supply chains, promoting innovative business by opening otherwise private information to help consumers to make better decisions (Thaler, 2013; Sunstein, 2012; Sayogo, 2013). Thus, considering the tradeoff between information confidentiality and privacy, and the benefits of opening data, a paradigm shift has already begun. This new paradigm encourages managers to securely open data to the public as part of investments or even as a revenue generating activity and not only an additional cost to the firm. This paradigm shift could change traditional mindset believing in direct causality between commercial confidentiality and profitability. In this context, first movers and leading firms will have to answer questions like: what data is truly confidential, and if breached, will it result in a significant loss to the organization? What data, if opened entirely, endangers the commercial privacy of the organization?

Once the true value of open data is perceived by the consumers, the expectations formed in the market also push companies to cope with the new paradigm. Thus, more and more organizations might discover that the release of previously restricted data might not only be harmless for their profitability but in fact create new economic benefits.

6.7.3 Business Process Models for Access Control Policies

A semantic web-based data platform such as the one described in Chap. 5 in this book will bring together heterogeneous actors and integrate disparate data sources. As argued in the section above, the heterogeneity in users and data sources exposes the information to more security risks. A system is as secure as its weakest link, and hence data extraction efforts by several heterogeneous actors can generate more vulnerability in the whole system. This situation calls for an automated intrusion detection mechanism due to the magnitude of possible information access requests and data extraction efforts.

Consequently, an access control mechanism should be designed and implemented to conduct authentication and authorization and govern access rights, as part of a formal business process (Fabian, Kunz, Müller, & Günther, 2013). Therefore, the various roles of employees in the organization and their different access rights required to accomplish their duties and also the patterns of the cooperation between employees should all be considered in designing and configuring this system. Due to these dimensions of complexity, it is clear that the access control mechanism should be designed as an integrated system capable of interchanging data with several subsystems, some of which may be based on different data structures and even in several formats.

However, opening data from one department might be detrimental to other functions in the organization. As discussed in Sect. 6.5.2, enough precautions should be employed to minimize a systematic inference attack on the disclosed data. Such attacks try to integrate data from different levels and several sources in order to discover some confidential information. When a user requests information about an organization's internal control system, the access control mechanism should be capable to determine the extent to which such information, in conjunction with other disclosed information, could potentially allow an inference attack to the accounting procedures in the organization. However, following such a procedure may create inefficiencies in the system. An alternative way of solving the problem consists of making an analysis during the time of designing the system and its corresponding data structure(s).

Integrating access control policies into business process requires managers to have an integrated perspective on their technology, people, and processes. Technology functions are backed by business processes, and business processes, in turn, support the aims of the policies. Behaviors supporting the policies are fostered by organizational rules and regulations to shape the organizational culture in the desired way. Consequently, values in open data initiative and also integrated access control mechanisms will both be parts of this business model.

6.7.4 Cooperation for Monitoring and Enforcement

As discussed above, the vast number of agents participating in semantic web platforms poses a challenge to maintaining information security, especially due to the assumption that each party has their own policies and protocols governing their information security. This challenge could hamper the initiative to create open data in certified coffee supply chain.

Consequently, we argue for the creation of a cooperative policy standard or an agreed upon framework for monitoring and enforcement. Procedures to govern different security policies and protocols will also be a part of such system. This framework can be developed under a new association whose members are the participating agents. This entity could serve three functions: assisting members in resolving conflicts emerging from different security policies and protocols, creating appropriate governance mechanisms for access control especially in terms of authentication and authorization of agents, and helping agents understand the impacts of different security policies and choose the best fit for their needs, while also maintaining the integration of the system as a whole at the desired level.

This proposition envisions the creation of standards that formalize the different information security policies, protocols, and procedures to enhance cooperative security policy for monitoring and enforcement. This standard will help harmonize the business rules in sustainable certified coffee with integrated or otherwise compatible information security policies and procedures. As the processes employed in the creation of certified coffee by various organizations involved in certified coffee supply chain are very similar, coming up with such a universal framework seems feasible. For instance, although the mix of the roasted coffee is a trade secret and entails commercial privacy value, the general process of transforming coffee bean into ground coffee is the same almost in all producers.

6.7.5 *Flexible Trust Management*

The availability of standards formalizing the different versions of information security policies could be useful in facilitating the creation of flexible trust management. The standard could facilitate a flexible policy framework for formulation of information policies. This framework would enable identification and negation of disparate policies and protocols for gaining access to information. Supported by the standard, the framework could be used to assess the balance between the acceptable trust level and risk level from each particular agent. The key tasks would be identifying the indicators and evidence necessary to induce trust. The issue in here is to find the right balance between efficiency of the indicators and the generality of the framework. Too general framework might need less stringent indicators, which could result in less efficient framework for policing the security of information. On the other hand, a too restrictive efficiency indicator in inducing trust might result in less generality in terms of the framework implementation.

6.8 Concluding Remarks

A global interoperable architecture based on semantic web technology that needs to access the data currently in the custody of private and public entities must be capable of maintaining information security. In an effort to protect the confidentiality and privacy of the data, custodians—private and public entities—should cooperatively enact security policies and procedures. These policies and procedures should govern the appropriate access level, minimize the inference attack, and provide a non-repudiation mechanism for the information that was legitimately disclosed. However, as a large number of users are expected in a semantic web platform, with heterogeneous data structures and security policies (O'Hara et al., 2004), concerns about protecting information as an economic and also public asset become crucial. Failing to build a trust in the effectiveness of a secured platform may become another barrier to actors participating in the platform (Fabian et al., 2013).

This chapter outlined challenges to information security and suggested that managers care about and come up with information security policies to mitigate the risks of unauthorized or illegitimate disclosure of information. This is especially important because it can affect the effective operation of semantic web platform for sustainable certified coffee. Challenges to information security of interoperable sustainable certified coffee supply chain platform stem can be categorized into at least three different classes[8]: incompleteness or error in the source(s) of data, complexity of the information environment, and fraud.

These challenges stem from the complexity of the information chains and the heterogeneity of stakeholders and data sources in the sustainable coffee supply chain. Achieving a platform that has an acceptable level of information security is only possible by integrating not only information security technologies (software/hardware products) but also people (awareness, skills, organizational culture) and business processes that are designed to back the other two dimensions in a way that satisfies security measures.

References

Buneman, P. (2013). The providence of provenance. In G. Gottlob, G. Grasso, D. Olteanu, & C. Schallhart (Eds.), *Big data* (Vol. 7968, pp. 7–12). Berlin, Germany/Heidelberg, NY: Springer. Retrieved from http://link.springer.com/10.1007/978-3-642-39467-6_3

Buneman, P., & Davidson, S. B. (2010, September 1). *Data provenance – The foundation of data quality*. Retrieved from http://www.sei.cmu.edu/measurement/research/upload/Davidson.pdf

Cavusoglu, H., Mishra, B., & Raghunathan, S. (2004). The effect of internet security breach announcements on market value: Capital market reactions for breached firms and internet security developers. *International Journal of Electronic Commerce, 9*(1), 69–104.

Choudhury, V., Hartzel, K. S., & Konsynski, B. R. (1998). Uses and consequences of electronic markets: An empirical investigation in the aircraft parts industry. *MIS Quarterly, 22*(4), 471–507.

Clemons, E. K., & Row, M. C. (1993). Limits to interfirm coordination through information technology: Results of a field study in consumer packaged goods distribution. *Journal of Management Information Systems, 10*(1), 73–95.

D'Aubeterre, F., Singh, R., & Iyer, L. (2008). A semantic approach to secure collaborative inter-organizational eBusiness processes. *Journal of the Association for Information Systems, 9*(3), 231–266.

[8] Based on the interviews with stakeholders in sustainable certified coffee and literary explorations.

Da Veiga, A., & Eloff, J. H. P. (2010). A framework and assessment instrument for information security culture. *Computers & Security, 29*(2), 196–207. http://doi.org/10.1016/j.cose.2009.09.002.

Fabian, B., Kunz, S., Müller, S., & Günther, O. (2013). Secure federation of semantic information services. *Decision Support Systems, 55*(1), 385–398. http://doi.org/10.1016/j.dss.2012.05.049.

Healy, P. M., & Palepu, K. G. (2001). Information asymmetry, corporate disclosure, and the capital markets: A review of the empirical disclosure literature. *Journal of Accounting and Economics, 31*(1–3), 405–440. http://doi.org/10.1016/S0165-4101(01)00018-0.

Huynh, T. D., Jennings, N. R., & Shadbolt, N. (2004). *Developing an integrated trust and reputation model for open multi-agent systems* (pp. 65–74). Presented at the 7th International Workshop on Trust in Agent Societies. Retrieved from http://eprints.soton.ac.uk/259557/

Jarman, H., Luna-Reyes, L. F., Zhang, J., Whitmore, A., Picazo-Vela, S., Andersen, D. L., et al. (2011). *I-Choose: Consumer choice, digital government, and sustainability in North America.* Presented at the APPAM Research Conference, Washington, DC.

Kagal, L., Finin, T., & Joshi, A. (2003). A policy based approach to security for the semantic web. In *International semantic web conference* (Vol. 2870, pp. 402–418). Sanibel Island, FL: Springer.

Klein, R., & Rai, A. (2009). Interfirm strategic information flows in logistics supply chain relationships. *MIS Quarterly, 33*(4), 735–762.

Leach, J. (2003). Improving user security behaviour. *Computers & Security, 22*(8), 685–692. http://doi.org/10.1016/S0167-4048(03)00007-5.

Lim, J. S., Chang, S., Maynard, S., & Ahmad, A. (2009). Exploring the relationship between organizational culture and information security culture. In *Australian information security management conference* (p. 12). Retrieved from http://ro.ecu.edu.au/cgi/viewcontent.cgi?article=1011&context=ism

Luna-Reyes, L. F., Zhang, J., Whitmore, A., Jarman, H., Picazo-Vela, S., Andersen, D. L., et al. (2014). Full information product pricing: An information strategy for harnessing consumer choice to create a more sustainable world. *Communications of the Association for Information Systems, 34*(1), 637–654.

Malhotra, A., Gosain, S., & Sawy, O. A. E. (2005). Absorptive capacity configurations in supply chains: Gearing for partner-enabled market knowledge creation. *MIS Quarterly, 29*(1), 145–187.

McLeod, A. J., Jr., Carpenter, D. R., & Clark, J. G. (2008). Measuring success in interorganizational information systems: A case study. *Communications of AIS, 22*(1), 617–634.

O'Hara, K., Alani, H., Kalfoglou, Y., & Shadbolt, N. (2004). *Trust strategies for the semantic web.* Presented at the Workshop on Trust, Security, and Reputation on the Semantic Web, 3rd International (ISWC'04). Retrieved from http://eprints.soton.ac.uk/260029/

Pahnila, S., Siponen, M., & Mahmood, A. (2007). Employees' behavior towards IS security policy compliance. In *40th Annual Hawaii International Conference on System Sciences, 2007. HICSS 2007* (p. 156b). http://doi.org/10.1109/HICSS.2007.206.

Panhuysen, S., & van Reenen, M. (2012). *Coffee barometer 2012.* Report, The Hague, Netherlands. Retrieved from http://www.teacoffeecocoa.org/tcc/Media/Files/Coffee-Barometer-2012

Rindfleisch, T. C. (1997). Privacy, information technology, and health care. *Communications of the ACM, 40*(8), 92–100. http://doi.org/10.1145/257874.257896.

Saint-Germain, R. (2005). Information security management best practice based on ISO/IEC 17799. *Information Management Journal, 39*(4), 60–66.

Sayogo, D. S. (2013). *Modeling incentives to disclose: Smart disclosure policy, private sector transparency and demanded disclosure.* Dissertation, University at Albany State University of New York, Albany, NY.

Sayogo, D. S., & Pardo, T. A. (2013). Exploring the determinants of scientific data sharing: Understanding the motivation to publish research data. *Government Information Quarterly, 30*(Supplement 1), S19–S31. http://doi.org/10.1016/j.giq.2012.06.011.

Smith, G. E., Watson, K. J., Baker, W. H., & Pokorski, J. A., II. (2007). A critical balance: Collaboration and security in the IT-enabled supply chain. *International Journal of Production Research, 45*(11), 2595–2613. http://doi.org/10.1080/00207540601020544.

Steinfield, C., Markus, M. L., & Wigand, R. T. (2011). Through a glass clearly: Standards, architecture, and process transparency in global supply chains. *Journal of Management Information Systems, 28*(2), 75–108. http://doi.org/10.2753/MIS0742-1222280204.
Stiglitz, J. E. (2000). The contributions of the economics of information to twentieth century economics. *The Quarterly Journal of Economics, 115*(4), 1441–1478. http://doi.org/10.1162/003355300555015.
Strong, D. M., Lee, Y. W., & Wang, R. Y. (1997). 10 potholes in the road to information quality. *Computer, 30*(8), 38–46.
Sunstein, C. (2012, March 30). *Informing consumers through smart disclosure*. Retrieved from http://www.whitehouse.gov/blog/2012/03/30/informing-consumers-through-smart-disclosure
Sutherland, W. J., Clout, M., Côté, I. M., Daszak, P., Depledge, M. H., Fellman, L., et al. (2010). A horizon scan of global conservation issues for 2010. *Trends in Ecology & Evolution, 25*(1), 1–7. http://doi.org/10.1016/j.tree.2009.10.003.
Tayi, G. K., & Ballou, D. P. (1998). Examining data quality. *Communications of the ACM, 41*(2), 54–57.
Thaler, R. H. (2013, May). *Smart disclosure: The regulatory future*. Presented at the Boulder Summer Conference on Consumer Financial Decision Making, Boulder, Colorado. Retrieved from http://leeds.colorado.edu/event/bouldersummerconference#2012conference
Thaler, R. H., & Sunstein, C. R. (2008). *Nudge: Improving decisions about health, wealth, and happiness*. New Haven, CT: Yale University Press.
Thaler, R. H., & Tucker, W. (2013, February). Smarter information, smarter consumers. *Harvard Business Review*. Retrieved from http://hbr.org/2013/01/smarter-information-smarter-consumers
Thompson, R. C. (2004). Lost at sea: Where is all the plastic? *Science, 304*(5672), 838. http://doi.org/10.1126/science.1094559.
Wang, R. Y. (1998). A product perspective on total data quality management. *Communications of the ACM, 41*(2), 58–65. http://doi.org/10.1145/269012.269022.
Wang, R., & Strong, D. (1996). Beyond accuracy: What data quality means to data consumers. *Journal of Management Information Systems, 12*(4), 5–33.
Zhang, C., & Li, S. (2006). Secure information sharing in internet-based supply chain management systems. *Journal of Computer Information Systems, 46*(4), 18–24.
Zimmerman, A. (2007). Not by metadata alone: The use of diverse forms of knowledge to locate data for reuse. *International Journal on Digital Libraries, 7*(1), 5–16.

Chapter 7
Long-Term Goals and Shifting Power Structures: A Convention-Based View

Francois Duhamel, Sergio Picazo-Vela, Isis Gutiérrez-Martínez, and Luis F. Luna-Reyes

Abstract Sharing information in supply chains may prompt conflicts of interest among stakeholders, presenting a challenge for achieving the long-term goals associated with platforms such as I-Choose. In this chapter, we analyze such potential conflicts and possible ways to overcome them, on the basis of convention theory and as a result of case studies. Through semistructured interviews with stakeholders of the coffee supply chain in the NAFTA region, we found the presence of four worlds, or "orders of worth": the domestic, civic, market, and industrial worlds, according to the terminology of convention theory. Our empirical work shows that in practice, supply chain participants can be characterized by a combination of at least two of such views. We also specify the conditions that make different supply chain configurations and set of values more or less amenable to the changes implied in the disclosure of private information that the I-Choose platform requires. In the conclusion of this chapter, we draw policy implications to design the right incentives to the private sector to enhance public value.

Keywords Certification • Coffee supply chain • Convention theory • Controversies • NAFTA

F. Duhamel (✉)
School of Business and Economics, Department of International Business, Universidad de las Americas Puebla, San Andres Cholula, Puebla 72810, Mexico
e-mail: francois.duhamel@udlap.mx

S. Picazo-Vela • I. Gutiérrez-Martínez
School of Business and Economics, Department of Business Administration, Universidad de las Américas Puebla, San Andres Cholula, Puebla 72810, Mexico
e-mail: sergio.picazo@udlap.mx; isis.gutierrez@udlap.mx

L.F. Luna-Reyes
Department of Informatics, University at Albany, Albany, NY, USA
e-mail: lluna-reyes@albany.edu

H. Jarman, L.F. Luna-Reyes (eds.), *Private Data and Public Value*, Public Administration and Information Technology 26,
DOI 10.1007/978-3-319-27823-0_7

7.1 Introduction

The first Fair Trade Seal—Max Havelaar—was created in 1988 as a result of a collaboration between the Mexican Cooperative UCIRI and the Dutch NGO Solidaridad (Fridell, 2007), with the aim of providing UCIRI producers with official recognition of their labor-friendly practices and thus providing them with access to new geographical markets. In 1993, several existing Fair Trade certification programs (including Max Havelaar) decided to group together under the umbrella of the Fairtrade Labeling Organization (FLO), creating an international network of Fair Trade organizations using the same certification standards. The FLO standards, as described in Chap. 4, are extensive and were historically applied exclusively to small farmers organized through cooperatives. However, in recent years, the annual 40 % average growth of the Fair Trade market in the US (Kim, Lee, & Park, 2010) has created strong pressures on the US Fair Trade organization to satisfy demand. In order to deal with those pressures, FLO's US partner, TransFair USA, decided to unilaterally extend fair trade certification to plantation and hired labor operations. For members of the traditional Fair Trade movement, certifying large-scale plantations represents a conflict of values given that many ancestors of large owners used to exploit the ancestors of current small producers. The lack of an agreement between TransFair USA and FLO on this issue resulted in the division of the Fair Trade movement into two main organizations, FLO and Fair Trade USA in 2011. Given that the US market is the largest coffee market in the world, the separation of Fair Trade USA gave rise to many changes in the Fair Trade ecosystem around the world.

The key question of our book, as stated in Chap. 1, is how to incentivize private actors to share their data in a way that promotes public value of the information disclosed. However, different organizations and consumers will not respond to the same incentives as they do not share the same set of values, thus creating a possibility of clashes even among organizations that share the same overarching objective. As illustrated above, TransFair USA's market-oriented values compelled them to create conditions designed to respond to the increased demand for fair trade coffee. This clashed with civic values represented by the FLO movement, which puts greater emphasis on maintaining the integrity of the original fair trade objective—the empowerment of small-scale producers. Both perspectives, each important and relevant in the Fair Trade movement, coexist along the certified coffee supply chain. This is to show that values, embedded in practice and carried out by stakeholders, play a fundamental role in the development of conflicts as well as facilitation of agreements within supply chains and certification systems, and can affect consensus on data disclosure.

Information-sharing platforms that are designed to enable information sharing across the supply chain are dependent for their success on the development of widely recognized and accepted information standards. Successful development of such standards requires engagement of stakeholders across the entire supply chain. Because stakeholders occupying different positions within the supply chain hold different value preferences, the process of standard creation might be fraught with conflicting priorities. As illustrated in Chap. 2, stakeholders in the coffee supply chain in NAFTA region hold different frameworks of reference that guide their definitions of quality and sustainability. The coexistence of those frameworks represents a source of further conflict, in addition to conflicts stemming from power imbalances in each supply chain governance mode. This poses specific challenges when introducing a platform like the one proposed in our project into an existing supply chain.

In this chapter, we use the lens of convention theory to address two main questions: what are the potential conflicts that can arise in the process of creating information-sharing platforms for the certified coffee supply chain and what conditions need to be created to overcome those conflicts? To answer these questions, we have organized this chapter into five sections, including this introduction. The following section consists of literature review that introduces convention theory as a lens for understanding potential sources of conflict in the certified coffee supply chain. The third section describes our research methods. The fourth section presents characterization of five types of supply chains based on their value preferences identified through an analysis of our interview data. The final section is used for a discussion of the different sources of conflict among the actors within the five supply chains and ways of resolving these conflicts. We seek clearer understanding of the sources of conflict to inform creation of governance mechanisms that might foster long-term adoption of information-sharing platforms through creation of adequate incentives to participate.

7.2 The Relevance of Convention Theories for Supply Chain Integration

Our main goal in this chapter, as we stated in the introduction, is to account for the possible conflicts that could impede efficient information sharing among the various actors in the certified coffee supply chains and to identify potential solutions to these conflicts. Effective information sharing is critical to platforms such as the one designed in our project, which cannot otherwise be sustained.

Institutional logics tend to explain conflicts within a given societal ecosystem as a consequence of the coexistence of different institutional orders (Gond & Leca, 2012). Thornton and Occasio defined institutional logics as "socially constructed, historical patterns of material practices, assumptions, values, beliefs, and rules by which individuals produce and reproduce their material subsistence, organize time and space, and provide meaning to their social reality" (Thornton & Ocasio, 1999, p. 804). Each institutional order has a central logic that guides its organizing principles and provides social actors with vocabularies of motive and a sense of identity (Thornton & Ocasio, 2008). Institutional logics theory has been used to describe how contending institutional orders, with different practices and beliefs, shape how individuals engage in political struggles (Friedland & Alford, 1991).

In this chapter, we apply the lens of convention theory, which is closely related to institutional logics (Thornton & Ocasio, 2008; Weber, Patel, & Heinze, 2013) although some basic assumptions of both approaches differ (Gond & Leca, 2012). The advantage of convention theory and its "economies of worth" approach (Boltanski & Thévenot, 2006) is that it considers the coexistence of multiple logics as inherent to ordinary social and organizational life, and not necessarily as an exception or a problem to be solved (Gond & Leca, 2012). In other visions of institutional logics, the most likely outcome of a struggle between distinct institutional orders would be an eventual dominance of one order over all others (Marquis & Lounsbury, 2007; Thornton & Ocasio, 1999). Our analysis of coffee supply chains aligns well with the convention theory, which directly addresses the patterns of coexistence of different definitions of product quality, or differing "orders of worth." It also identifies theoretical approaches to solving conflicts stemming from the presence of conflicting values, specifying how people view the world in general and the legitimation of their activities in particular.

Convention theory recognizes that there is no "objective" definition of product quality expressed exclusively by differences in prices in the market. Quality standards depend on the shared identification of common characteristics that can be grouped into three categories: *search attributes,* which can be verified at the time of the transaction; *experience attributes,* which can be assessed only after the transaction has taken place; and *credence attributes*, which cannot be objectively verified and are based on trust (Darby & Karni, 1973; Nelson, 1970). A product represents not only the outcome of a material production process, but is at the center of a web of contractors, distributors, consumers, and regulators, which develops over time in a path-dependent manner driven by taken-for-granted assumptions, common practices, and shared conventions that define quality and maintain stable relations (Biggart & Beamish, 2003). Conventions in that context can be defined as "shared templates for interpreting situations and planning courses of action in mutually comprehensive ways that involve social accountability" (Biggart & Beamish, 2003, p. 444). Conventions appear as "models of evaluation" where actors need to agree on the attributes associated with a given definition of product quality. The tools that are consequently used to ensure such quality depend for their legitimacy on shared values among the actors (Ponte & Gibbon, 2005). Assumptions and values involved in the development of these conventions can be conceptually classified in several "worlds" that specify how reality is grasped by different constituencies, and "the format of what constitutes information" (Thévenot, 2006, 2007), and how this information is communicated.

Conventions serve several functions: they express actors' values, help actors justify their actions in a particular world, and enable coordination of behavior (Eymard-Duvernay, 1989). The convention theory identifies four worlds, or "orders of worth": market, domestic, industrial, and civic. The coordination mechanism of the market world is price information, which expresses worth and serves to justify actions in this world. Domestic worth qualification places value on experience and seniority, and legitimacy of information is based mainly on trust. Industrial worth qualification values operational expertise in the efficient execution of a task, with legitimacy depending on consensus about the correctness of the technique employed. Civic worth qualification emphasizes social values, and legitimacy is based on respect of law and social relevance (Boltanski & Thévenot, 2006; Ponte & Gibbon, 2005).

In addition to defining the coordination mechanisms of the different orders of worth, convention theory allows us to identify sources of potential conflicts between actors that exhibit different orders of worth. For example, people in the domestic order tend not to like the anonymity of the civic world, the corruption of market relations, and the unnecessary formalism and standardization of the industrial world. The civic order tends not to appreciate the dependence on personal relations of the domestic world, which are seen as leading to particularism, paternalism, and corruption. Civic order may consider market coordination with suspicion for its individualism and insistence on particular interests and selfishness. The civic order sees the industrial world as dominated by unnecessary technocracy and bureaucracy. Market order, on the other hand, promotes liberalization from the domestic world, through abolishment of personal links, particularism, and personal prejudices to access a borderless, anonymous world. Market order does not go well with the public space promoted actively by the civic order. People in the market order prefer contracts and face-to-face relations rather than open justice in the public space. Market order also criticizes the lack of flexibility of industrial order's tools, methods, and structures. For the industrial world, the domestic world is judged as traditional and outdated. Its particularism is judged inefficient and superiors relying on authoritarianism are judged incompetent. Administrative procedures are considered excessive and social policies unnecessarily expensive in the industrial world, which also resents lavish consumption, high prices, and the fluctuations in prices of the market order (Ponte & Gibbon, 2005).

The conflicting values of the different orders of worth are an obvious source of conflict in social institutions where actors with different values coexist. In order to maintain functionality of such institutions or systems, actors must find ways to resolve these conflicts. Boltanski and Thévenot (2006) identify three potential outcomes for conflicts between orders of worth: agreement, compromise, and relativization. Agreement is an instance in which conflicts within the same order of worth can appeal to superior common values to arbitrate disagreements. The need for compromise arises when disagreements occur between orders of worth. In such situations, appealing to a superior common value is not feasible as actors inherently disagree on the importance of basic values and justify their actions according to different orders of legitimacy. Relativization occurs when actors agree to find a compromise and conclude a private agreement without resorting to debate over principles: "A private arrangement is a contingent agreement between two parties that refers to their mutual satisfaction rather than to a general good (*you do this which is good for me; I do that, which is good for you*)" (Boltanski & Thévenot, 2006, p. 336).

Convention theory helps with formulating conditions for overcoming potential conflicts in the transmission of information and refining policy recommendations. However, the way actors try to solve conflicts within and across organizations where

Table 7.1 Sources of conflict and compromise between civic order and market, industrial, and domestic orders

	With market order of worth	With industrial order of worth	With domestic order of worth
Source of conflict with civic order of worth	Suspicion stemming from alleged manipulation of information asymmetries, due to protection of particular corporate interests	Resented for its focus on technology and measurability	Resented for the secrecy of personal relations
Source of compromise at the value level with the civic order of worth	Aligning civic values with market values through adoption of civic values by consumers	Reintroducing social rights to favor productivity, against waste	Aligning civic values with personalized relations
		Integration of technology	

Source: Own elaboration from Boltanski and Thevenot (2006) and Patriotta et al. (2011)

a plurality of normative orders prevails has been studied only sparingly in the present literature (Gond & Leca, 2012; Patriotta, Gond, & Schultz, 2011). Some of the existing literature focuses on the possible compromises that can be found between different orders of worth, such as the increasing adoption of selected norms from one another (Ponte & Gibbon, 2005), which has resulted in a degree of value interpenetration. For example, industrial and market orders of worth have worked together using industrial norms of productivity, economies of scale, and technical progress. Also, market order of worth has adopted some forms of domestic coordination when marketing of a branded product is based on geographic location. In market coordination, people may work together with civic partners to accommodate products fulfilling a series of minimal norms. Compromise between market and domestic orders is more difficult to achieve as we will show later on in our interview analysis. We illustrate some of these potential sources of conflict and compromise in Table 7.1.

Convention theory can be criticized for overlooking institutional issues linked to power and domination. For example, some actors may not have the capacity to criticize or contest dominant social orders, while others may access multiple logics due to their position at the intersection of different orders of worth and can use the different logics to justify and impose their views (Greenwood, Díaz, Li, & Lorente, 2010). Integrating insights of value chain governance theory (Gereffi, Humphrey, & Sturgeon, 2005) into convention theory, as done by Ponte and Gibbon (2005), addresses this objection. Value chain governance theory takes into account three main variables when considering coordination strategies between actors: (1) the complexity of information exchanged, (2) the "codifiability" of information exchanged, and (3) the capabilities of the supplier base. Complexity refers to the volume of non-price information flowing across interfirm boundaries, "codifiability" refers to the extent to which information and knowledge needed for conducting transactions can be codified and transmitted efficiently, and capabilities of the supplier base refer to the capabilities of suppliers to respond to control and monitoring requirements (Gereffi et al., 2005). This is relevant to our research because every supply chain configuration exhibits a lead organization that determines who does what along the chain, at what price, using which standards, and under which specifications (Muradian & Pelupessy, 2005; Ponte & Gibbon, 2005). The analysis of a supply chain reveals not only contrasting world views but also power relations between partners in a chain, conditioning the possibilities of transmission of information and the possibilities of adoption of technical standards that codify information.

We found two studies applying convention theory to the coffee sector. In the first one, Ponte and Gibbon (2005) identify three main orders of worth: the domestic world, dominated mostly by producers and niche marketers of specialty coffee; the civic world, dominated by marketers of ethical products selling fair trade, organic, and other sustainable coffee; and finally, a combination of two worlds, the industrial-market world, where retailers sell branded, mainstream coffee. In line with their work, we consider the presence of all four orders of worth but treat industrial and market orders as separate for the purposes of our analysis. The second study focuses on the difficulty of harmonizing civic and market values around fair trade coffee, because of a "contradiction between the identity of the groups linked to activism and their reality as business" (Renard, 2003, p. 92). This contradiction results in the division of fair trade into two streams: one reabsorbed by market forces and the other at the service of "alternative" producers.

Although convention theory addresses standard setting, the role of certifiers and certification processes has not yet been thoroughly analyzed under that lens, despite the important role certifiers play in the certified coffee market. Certifiers are key to the definition and effective transmission of quality standards in supply chains, not only at the service of large retailers but also at the level of smaller producers. In doing so, they influence the reconfiguration of global supply chains. Hatanaka, Bain, and Busch (2005) contend that third party certification supports alternative practices of small producers but has also become a tool used by dominant retailers to increase their market power in global agrifood chains. At the same time, Reardon, Codron, Busch, Bingen, and Harris (2000) observe that large firms and multinationals in the global agrifood sector often create private grades and standards, which is also the case with large specialty retailers such as Starbucks (Macdonald, 2007). These powerful actors drive other actors, such as smaller domestic firms and farms in emerging markets, into adopting comparable standards to gain access to desirable export markets. Some governments may develop programs to help smaller producers invest in upgrading their farms according to the requirements demanded by the dominant grades and standards. The smaller firms may also choose to ally with public and non-profit sectors to establish their own grades and standards (Reardon et al., 2000).

Convention theory deals with observing different world visions of stakeholders (consumers, distributors, roasters, and producers in the case of coffee) and how they enter into conflict or harmony to bring about a negotiated order of information transmission. The task of the researcher is to compare the plurality of legitimate orders of worth and look at the needed compromises between stakeholders (Thévenot, 2006) in respect to information formats and transmission of reliable and up-to-date data to overcome conflicts. To address those issues, we conducted a series of interviews with participants in the coffee supply chain as we describe in the following section and used this data to draw theoretical and empirical implications that are presented in the discussion and conclusion sections.

7.3 Methods

In order to answer our research questions, we interviewed multiple stakeholders along the coffee supply chain following a multiple case study approach (Yin, 1994), in which analysis of individual cases is followed by an analysis aimed at comparing and contrasting their similarities and differences. To identify participants for the study, we followed a snowball sample approach, starting with interviewees from coffee associations who were then asked to identify other potential candidates. This process was reiterated with each consecutive interview. The semistructured interview protocol included questions about information sharing, conventions on quality, characteristics of an information standard, motivations, barriers, and potential conflicts to share information across the supply chain. Each interview lasted approximately 1 h.

We conducted 25 interviews from September to December 2013. We interviewed three producers, three intermediaries, one roaster, and seven retailers. In addition, we interviewed some participants that were involved in more than one process: two that were intermediaries and roasters and two that were producers, intermediaries, roasters, and retailers. We also interviewed two members of a coffee association and five certifiers. All interviews were recorded and transcribed with the exception of one where interviewee did not give permission for recording. In this case, three interviewers documented the interview.

To compare and contrast responses from participants, we coded all the interviews. We worked together to define a set of 13 categories including themes such as barriers, motivations, values, conflicts, and quality. We coded each interview according to these themes, and each comment was classified as representative of one of the four worlds. That is to say, we classified comments in terms of the worldview they represented: civic, market, industrial, or domestic. In this way, by analyzing the interviews, we identified five different types of supply chains. We arranged codification tables belonging to each type of supply chain to understand the perspective of their members. Finally, we compared and contrasted the five perspectives. Although we are reporting the results in English, data collection and analysis were done in Spanish by native speakers.

7.4 Results

As mentioned in the previous section, our analysis of the interview data led to characterization of five different types of coffee supply chains. Each of the five supply chains works in different ways, with each following different definitions of coffee

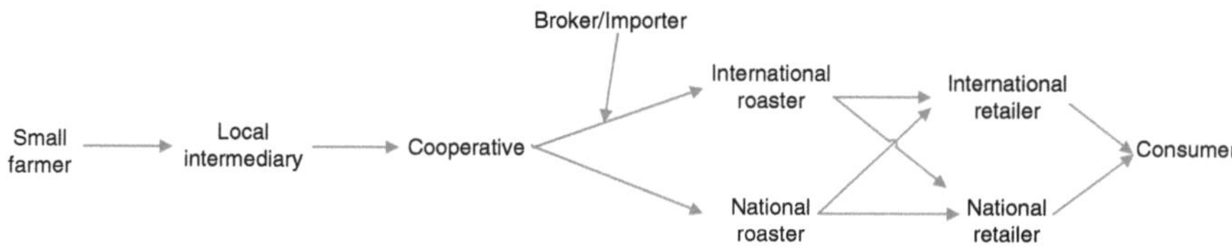

Fig. 7.1 The traditional fair trade supply chain

quality. What we found is that each supply chain exhibited adherence to more than one of the orders of worth defined by convention theory, with many actors valuing elements of all four worlds. In order to classify the five identified supplied chains, we used the two dominant orders of worth present in each of the chains to describe their actors' guiding value set. Based on this approach we identified the following coffee supply chains:

7.4.1 Traditional Fair Trade Supply Chain

The first type of supply chain represents what we have called the traditional fair trade supply chain (see Fig. 7.1). In this supply chain, small farmers organized in local cooperatives sell their coffee to the Mexican and the international market, usually in the form of green coffee beans. The farmers who own small plots of land pick the cherries and remove the pulp around the coffee beans in a process usually known as the "wet process." In order to transport the finished product to the main processing site, coffee beans are gathered by a local intermediary, usually a local representative of the cooperative. Coffee beans are then moved to the central cooperative location, where they undergo a second "dry process" before being sent to the roaster. Certified coffee is usually sold to the international market, while the noncertified coffee also produced by cooperative members is sold to the national market. Roasters then distribute the coffee to the consumer using a retailer, which in some cases is the roaster himself.

The *civic* and the *industrial* orders of worth are present in the traditional fair trade supply chain, with the civic order being the dominant order. Participants in this supply chain frequently refer to the importance of establishing a fair relationship between producers and consumers. For example, one representative of a Mexican cooperative commented "the objective of Fair Trade is – as the name suggests – to establish a fair exchange between producers and consumers… Working together… the product is supervised by experts who certify that the products are fair and comply with specific processes, and the consumer responds to our effort by paying a fair price for our products."

Conversations with participants in this supply chain are also full of references to quality of process, products, and compliance to certification, which are closely related to the industrial order of worth. For example, one of the managers of a local trading organization explained their mission as "Our organization serves about

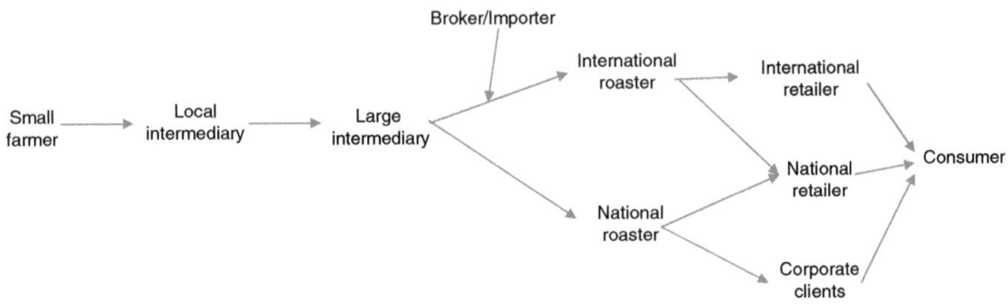

Fig. 7.2 The "mixed" supply chain

60 producers in Mexico. We are not a trading company in the strict sense because we are technical staff supporting product development and process improvement to help our producers comply with quality, safety and other legal requirements." Another example of the combination of the *civic* and *industrial* orders of worth can be found in the fair trade certification processes, which consist of norms and standards (representing the *industrial* world) that define product quality in terms of social, environmental, and economic factors (representing the *civic* world).

7.4.2 Mixed Supply Chain

The second type of supply chain is a "mixed" supply chain (see Fig. 7.2). There are two subclasses encompassed within this classification titled type I and type II mixed supply chain, which involve the same participants but have different modes of operation. The chain begins with small independent farmers not organized in a local cooperative, who pick the cherries and run the "wet" process before selling the coffee beans to local intermediaries. Local intermediaries sell the coffee to large intermediaries who run the dry process before selling the coffee beans to either national or international roasters. Roasters in turn sell the coffee to the consumer through retailers or corporate clients.

The first mode of operation of this supply chain (type I mixed) follows a logic, which is a mix of the *industrial* and *civic* worlds (where industrial order of worth is dominant over civic order of worth), and we found it mainly in the Nestlé supply chain. The second mode of operation (type II mixed) is a mix of the *industrial* and *market* world views, and it is represented by other Mexican roasters that sell their products both in the national and international markets.

As mentioned above, the type I mixed supply chain (industrial-civic) is illustrated by the Nestlé mode of operation. The Nestlé supply chain is currently working under the principles of the 4C certification program on a global initiative called the "Nescafe Plan." The main objective of the Nescafe Plan is to provide traceability to each jar of Nescafe to enable the consumer to trace the origins of the particular coffee she is about to consume. The plan also involves a series of interventions to improve production and processing practices to make the products more environmentally, socially, and economically sustainable. The focus on traceability, process improvement, and quality control are the main references to the industrial world, and the sustainability concerns are closely related to the civic world. Both views are continuously mentioned in the interviews with participants in this supply chain. For instance, one of the large intermediaries selling coffee beans to Nestlé commented that "something that coffee requires from the plant to your cup is quality control. If you do not have quality control in every step, even when you are brewing it, the coffee will be spoiled." He also highlighted the importance of adhering to the specific processes and improving sustainability as a necessary condition in order to remain part of the Nestlé program, "Nestlé is a very innovative and trustworthy company to us, and a very attractive client. They are always at the cutting edge. For example, nowadays, if you want to sell coffee beans to Nestlé, you are required to have a sustainability certification that involves food safety as well as environmental sustainability." This view is also shared by Nestlé representatives. One of them described the program as "a 500 million Swiss Francs with a holistic approach to promote sustainability and sustainable consumption. Currently there are 22 million 4C coffee plants, which comply with social, economic and environmental principles coming from the Rainforest Alliance and UTZ."

Although this supply chain emphasizes the same basic principles that motivate the traditional Fair Trade coffee supply chain, the main trigger for the movement is not on the side of the producer, looking for a more fair treatment, but on the side of the consumers and their information needs. For example, when we asked an intermediary about information that could be useful to him, he commented that "it is more about the information that consumers want to know, which is a first-world trend, and these needs go hand-in-hand with the main benefits to the environment that Nestlé is looking for." Nestlé representative also stressed the idea of the consumer as the main driver. For example, when asked about definitions of quality, one Nestlé marketing representative described a quality coffee as "the coffee that consumers like, and once we know what they want, it is our commitment to always provide the same flavor and quality."

The type II mixed form of operation in this type of supply chain was most committed to the *market* and *industrial* worlds, where market order of worth is dominant over industrial order of worth. The quality of coffee and the process needed to ensure it were topics continuously mentioned during the interviews. The interviewees from the type II mixed supply chain mainly made references to market conditions and price, with transparency and sustainability being completely omitted. For example, when describing the perception of the company about organic and other certificates, a representative of an important Mexican roaster commented: "I do not see a lot of value in that [certifications]. We have a Kosher seal that goes to a very specific market segment, and an organic coffee that people do not buy much. I mean, I do not see these seals as a competitive advantage because of the price. I think that people [in Mexico] do not decide to buy a coffee like that." In general, interviewees involved in supply chains that mostly produce coffee for the Mexican

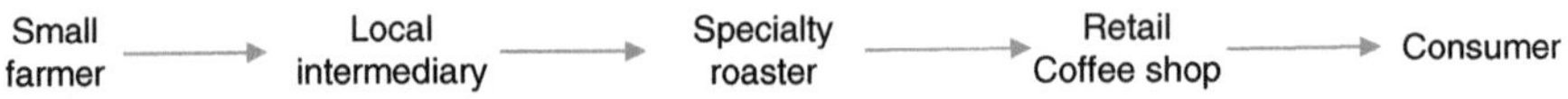

Fig. 7.3 The specialty coffee or relational supply chain

market were far less interested in certifications than those exporting a large portion of their products.

7.4.3 Specialty or the Relational Supply Chain

The third type of supply chain can be characterized as the specialty or relational supply chain (see Fig. 7.3). In this supply chain, local intermediaries are usually recognized coffee cuppers who build close relationships with both producers and roasters. The cupper helps producers improve their practices to gradually increase the quality of their coffee beans and helps the roaster improve roasting techniques to create specialty coffees that are usually sold in retail coffee shops, each of them featuring a specific mix of coffees that make the experience at the coffee shop unique. Interviews with participants in this supply chain made us infer that the main values in the supply chain were the *industrial* and the *domestic* orders of worth, with the industrial order being dominant. Conversations with participants in this supply chain emphasize the quality of the coffee and the importance of quality control processes along the entire production chain, signifying the dominance of the industrial order of worth. The domestic order of worth is illustrated by their emphasis on the importance of promoting Mexican producers in Mexico. Moreover, the domestic order of worth is also demonstrated by the value specialty coffee shops we interviewed placed on the relationships with their producers. The interviewees emphasized that for them defining coffee quality comes from the quality of the relationships within the supply chain, rather than from external certifications.

7.4.4 Medium-Sized Farmer Supply Chain

The last type of supply chain that we found during our data gathering process involves small- and medium-sized enterprises (SMEs), in some cases, family businesses, who own a medium-sized plantation or several of them (see Fig. 7.4). In general, these supply chains were more hierarchically organized, enveloping the entire production process from growing to roasting under a single brand. Occasionally, these ventures have to supplement their production by purchasing coffee from local intermediaries to be able to satisfy demand for their product. Although not all interviewed SMEs had a sustainability certification, all SMEs that exported their product to foreign markets had a certification, recognizing the price premium they could receive from the seal. The medium-sized producers get to the

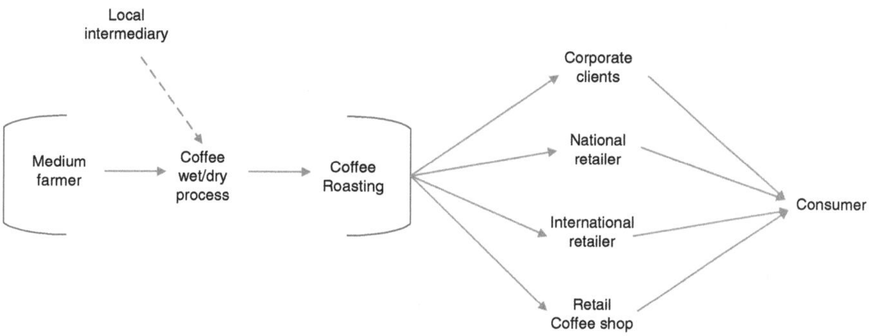

Fig. 7.4 The medium-sized farmer supply chain

end consumer in many different ways, including international retailers, national retailers, corporate clients, or even their own coffee shops. We found that participants in this type of supply chain mainly exhibit values compatible with the *industrial* and *domestic* worlds, with industrial order of worth being dominant. The industrial commitment comes from their emphasis on the production process, and the quality control from the selection of the beans to the roasting process. Because those are local or family businesses, they also emphasize personal relations and trust in doing business, which makes them part of the domestic world.

It is important to mention some other relevant results that come from our observations and interview data. First, small farmers have no access to certifications because they do not have the resources or the technical knowledge to go through a certification process. Certifications are generally awarded to cooperatives, large intermediaries, and medium farmers, who have the resources to undergo the necessary certification processes and obtain a certification. In this way, any relationship to gather information from small producers—who produce about 80 % of all coffee in Mexico—needs to be mediated by these intermediaries.

Second, we found that participants in different types of supply chains have conflicts that are driven by historical context and thus may be hard to address. For example, medium-sized farmers are in many cases family businesses that in the past were owners of big plantations that exploited Mexican Indians. Although most of them are currently concerned about being fair to their workers, cooperative members and fair trade advocates in Mexico do not agree with allowing them to be Fair Trade certified. Curiously enough, they provide a market reason for that. For example, one cooperative representative mentioned that "a small producer has only few spaces to get into the market, and fair trade certification is a way to protect that market."

Finally, large corporations, such as Nestlé, have the recognition, even admiration, from medium-sized farmers and also from other large participants in the supply chains. On the other hand, small farmers and fair trade advocates are very suspicious about their current practices. This conflict between large and small producers is at the core of the division of the global Fair Trade movement and represents a major risk for any project that aims to create information-sharing platform across the entire certified coffee ecosystem.

7.5 Discussion and Concluding Remarks

As described in the previous section, four potentially conflicting orders of worth interact in the coffee supply chain: the civic world, the industrial world, the domestic world, and the market world. In our research, we identified five different types of supply chains classified by different combinations of exhibited values: civic-industrial (the "traditional fair trade" supply chain), industrial-civic (the "type I mixed" supply chain), industrial-market (the "type II mixed" supply chain), industrial-domestic (the specialty coffee or relational supply chain), and the domestic-industrial (the "medium-sized famer" supply chain). The presence of mixed motivations in the five identified supply chains challenges the depiction of the role of values in the convention theory, which treats the four orders of worth as ideal types. We contend that a singular actor or singular supply chain can exhibit more than one order of worth as long as the underlying values can be reconciled. In fact, we observed that inside the same organization, individual departments had different "orders of worth" depending on their proximity to production practices, quality control, or to the final consumer market. These combinations define types of institutional arrangements that might oppose each other but also may coexist in an ecosystem catering to different needs.

Four worlds, or orders of worth, seem to be at play in the coffee industry: *Civic* values are becoming increasingly important in the negotiation of the definition of coffee quality (paying a fair price, helping small farmers' organizations). At the same time, labeling and certification systems are organized in terms of an *industrial* convention, and relationships with some mainstream distributors who carry fair trade coffee are based on a *market* convention. It is interesting to note the appreciation of all the interviewees for the values in the industrial world, which may be a key point to facilitate any long-term relationships in a platform such as I-Choose. Our interviews also suggest that the main potential source of conflict in the coffee supply chain comes from the contradicting values in the market and civic worlds, which already caused a division in the Fair Trade movement.

"Lead firms," such as Nestlé in type I mixed supply chains or other big firms in type II mixed supply chains, define and manage "quality" by shaping the rules and conditions of participation and determining the functional division of labor along the chain, sometimes with support of certification programs such as 4C. Such leading firms diffuse dominant normative paradigms that provide legitimacy for the mechanisms used to exert "leadership." The actual forms of coordination between lead firms and first-tier suppliers (and their hands-on or hands-off character) vary depending on (1) the mechanisms for transmitting knowledge and information about quality and (2) the values guiding the lead firms. Their success depends on how well they transfer information to their suppliers and to standardize, codify, and obtain credible external certification.

In specialty coffee markets, more information is provided about the coffee's origin and its environmental and other impacts. In these markets, actors tend to use narratives such as origin-based trust narratives. These narratives are being increasingly replaced by certified quality systems such as the coffee standards developed by the Specialty Coffee Association of America that partially dissociate coffee's quality from its place of origin. These narratives tend to be replicated in a standardized manner for mass consumption (Starbucks), thus recalling industrial quality conventions.

The system architecture introduced in our information-sharing platform carries civic values of transparency, favoring public goods, and public services for the general public and consumers. Such system can be read as an attempt to introduce civic order of worth into an environment normally dominated by industrial/market order of worth, and thus be classified along the lines of a traditional fair trade supply chain that exhibits values of civic-industrial order of worth. This leaves us with two questions: What kind of conflicts might we expect as a result of introducing a system-wide information-sharing platform? And how can such conflicts be overcome? We expect that the resulting conflicts will need to be resolved via compromises, and not as much through agreements and relativization of conflicts as described by Boltanski and Thévenot (2006). In Table 7.2, we describe the various sources of conflict between the civic-industrial type of a supply chain as a stand-in for an information-sharing platform and the other four types of supply chains identified through our empirical exploration as well as potential sources of compromise.

As we show in Table 7.2, the traditional fair trade supply chain and the industrial-civic mixed supply chain are in agreement about core values. However, they are not necessarily in agreement about the means to support such values. For

Table 7.2 Potential points of friction and sources of compromise with I-Choose

	1	2	3	4
Order of worth	Industrial-civic	Industrial-market	Industrial-domestic	Domestic-industrial
Type	"Type I mixed"	"Type II mixed"	"Specialty coffee"	"Medium-sized famer"
Source of conflict with civic-industrial order of worth	Agreement in the ends but conflicts about the means	Protection of corporate interests vs. producer and consumer well-being	Trust in quality comes from personal relationships rather than certification systems	Domestic values are closely related to family businesses that used to be owners of big plantations
Source of compromise with the civic-industrial order of worth	Harmonizing ways of being fair with the producer	Consumer interest in civic values may align market and civic worlds	Aligning people and product certifications	Negotiating market segments and differentiation

example, Nestlé program involves supporting high-quality producers with a monetary prize, which is independent from the price paid for the coffee. Traditional fair trade farmers see this monetary price as a low-impact intervention. Moreover, traditional fair trade values self-improvement rather than external interventions to improve the quality of life of the small farmers. Compromises then can be made by harmonizing the means of fair relationships between main actors in both supply chains, which should include small farmer involvement in the conversations. Currently, most conversations and certifications are mediated in these mixed supply chains by intermediaries.

Market focus of the type II mixed supply chain conflicts with the civic interests of the traditional fair trade supply chain because market prices do not include externalities from the production and distribution process. Participants in this type of supply chain showed little interest in the civic values mainly because there is not yet consumer interest in paying the price premium that implies a more sustainable commerce. Promoting market penetration of fair trade and other sustainable products is a key source of compromise with the market view. In a sense, type I mixed supply chains are, from our point of view, supply chains that have already started the transition process given the projected demands. Corporations in type I mixed supply chains are the leaders in the market, and we believe that other corporations will follow once consumers become more aware of civic values.

Specialty coffee supply chains do not emphasize product certifications because the main source of trust in the quality of the coffee comes from personal relationships among supply chain participants. We had a chance to interview some of the participants in this kind of supply chain who did not know that the coffee that they were selling was actually certified following international norms. In their view, the quality of the coffee was attached to the coffee cupper behind the process. Coffee cuppers are also recognized by personal certifications, and we believe that including these personal credentials as an alternative source of trust may be a way of reaching compromise with participants in this supply chain.

Finally, the conflict that might be the most difficult to resolve is between participants in the medium-sized farmers supply chain, which exhibits domestic-industrial values, and participants in the traditional fair trade supply chain. The main source of conflict stems from the fact that the domestic order of worth among this type of supply chain is rooted in the importance of family relations, which inevitably brings in the legacy of exploitation of small farmers and Indians. Coffee producers in this supply chain are usually descendants of landlords who exploited Indians and other small farmers in the past, which creates a historical division between both groups. Giving medium-sized farmers an opportunity to enter the fair trade market is interpreted as an inconsistency with the traditional values of fair trade, because of this historic divide. One way to look for compromise might be to develop alternative seals and certifications that clearly differentiate both producers and may reach different market segments.

In order to enhance public value, the findings from the interviews suggest that the accomplishment of compromise between actors belonging to different orders of worth and types of supply chains is necessary to incentivize participation of private actors in information-sharing platforms. In these circumstances, public administration can play an important role by creating the environment for the establishment of alternative certifications and by demonstrating the benefits of sharing information. They should help to overcome potential barriers created by mistrust among companies (see Chap. 1). Moreover, policy makers should pay special attention to the diversity of values in the coffee supply chain which lead to the need for collaborative governance like the one that we will describe in the next chapter of the book.

References

Biggart, N. W., & Beamish, T. D. (2003). The economic sociology of conventions: Habit, custom, practice, and routine in market order. *Annual Review of Sociology, 29*(1), 443–464. http://doi.org/10.1146/annurev.soc.29.010202.100051.

Boltanski, L., & Thévenot, L. (2006). *On justification: Economies of worth.* (Catherine Porter, Trans.). Princeton, NJ: Princeton University Press.

Darby, M. R., & Karni, E. (1973). Free competition and the optimal amount of fraud. *Journal of Law and Economics, 16*(1), 67–88.

Eymard-Duvernay, F. (1989). Conventions de qualité et formes de coordination. *Revue Économique, 40*(2), 329–359.

Fridell, G. (2007). *Fair trade coffee: The prospects and pitfalls of market-driven social justice* (Repr.). Toronto, ON: University of Toronto Press.

Friedland, R., & Alford, R. R. (1991). Bringing society back in: Symbols, practices and institutional contradictions. In W. W. Powell & P. DiMaggio (Eds.), *The new institutionalism in organizational analysis* (pp. 232–263). Chicago: University of Chicago Press.

Gereffi, G., Humphrey, J., & Sturgeon, T. (2005). The governance of global value chains. *Review of International Political Economy, 12*(1), 78–104.

Gond, J. P., & Leca, B. (2012). Theorizing change in a pluralistic institutional context: What can economies of worth and new-institutionalism learn from each other? *Lille Economie & Management-LEM*, Working Paper no. 15, 53 p.

Greenwood, R., Díaz, A. M., Li, S. X., & Lorente, J. C. (2010). The multiplicity of institutional logics and the heterogeneity of organizational responses. *Organization Science, 21*(2), 521–539. http://doi.org/10.1287/orsc.1090.0453.

Hatanaka, M., Bain, C., & Busch, L. (2005). Third-party certification in the global agrifood system. *Food Policy, 30*(3), 354–369. http://doi.org/10.1016/j.foodpol.2005.05.006.

Kim, G.-S., Lee, G. Y., & Park, K. (2010). A cross-national investigation on how ethical consumers build loyalty toward fair trade brands. *Journal of Business Ethics, 96*(4), 589–611. http://doi.org/10.1007/s10551-010-0486-1.

Macdonald, K. (2007). Globalising justice within coffee supply chains? Fair trade, Starbucks and the transformation of supply chain governance. *Third World Quarterly, 28*(4), 793–812. http://doi.org/10.1080/01436590701336663.

Marquis, C., & Lounsbury, M. (2007). Vive La Résistance: Competing logics and the consolidation of U.S. community banking. *Academy of Management Journal, 50*(4), 799–820. http://doi.org/10.5465/AMJ.2007.26279172.

Muradian, R., & Pelupessy, W. (2005). Governing the coffee chain: The role of voluntary regulatory systems. *World Development, 33*(12), 2029–2044. http://doi.org/10.1016/j.worlddev.2005.06.007.

Nelson, P. (1970). Information and consumer behavior. *Journal of Political Economy, 78*(2), 311–329.

Patriotta, G., Gond, J.-P., & Schultz, F. (2011). Maintaining legitimacy: Controversies, orders of worth, and public justifications: Maintaining legitimacy. *Journal of Management Studies, 48*(8), 1804–1836. http://doi.org/10.1111/j.1467-6486.2010.00990.x.

Ponte, S., & Gibbon, P. (2005). Quality standards, conventions and the governance of global value chains. *Economy and Society, 34*(1), 1–31. http://doi.org/10.1080/0308514042000329315.

Reardon, T., Codron, J. M., Busch, L., Bingen, J., & Harris, C. (2000). Global change in agrifood grades and standards: Agribusiness strategic responses in developing countries. *The International Food and Agribusiness Management Review, 2*(3), 421–435.

Renard, M.-C. (2003). Fair trade: Quality, market and conventions. *Journal of Rural Studies, 19*(1), 87–96. http://doi.org/10.1016/S0743-0167(02)00051-7.

Thévenot, L. (2006, July). *Institutions and agency: Differentiating regimes of engagement*. Presented at the Conference on Economy and Society, Italy.

Thévenot, L. (2007). The plurality of cognitive formats and engagements moving between the familiar and the public. *European Journal of Social Theory, 10*(3), 409–423.

Thornton, P. H., & Ocasio, W. (1999). Institutional logics and the historical contingency of power in organizations: Executive succession in the higher education publishing industry, 1958–1990. *American Journal of Sociology, 105*(3), 801–843. http://doi.org/10.1086/210361.

Thornton, P. H., & Ocasio, W. (2008). Institutional logics. In R. Greenwood, K. Sahlin, & R. Suddaby (Eds.), *The SAGE handbook of organizational institutionalism* (pp. 99–129). Los Angeles: Sage.

Weber, K., Patel, H., & Heinze, K. L. (2013). From cultural repertoires to institutional logics: A content-analytic method. In M. Lounsbury & E. Boxenbaum (Eds.), *Institutional logics in action, part B* (Vol. 39, pp. 351–382). Bingley, UK: Emerald Group Publishing Limited.

Yin, R. (1994). *Case study research: Design and methods*. Newbury Park, CA: Sage.

Chapter 8
Encouraging Private Sector Transparency: Policies to Support Disclosure of Product Data in North America

Holly Jarman, Luis F. Luna-Reyes, and Theresa A. Pardo

Abstract Our concluding chapter draws on the concepts and theories discussed in the book, particularly the concept of governance, collaboration, and the role of trust. The chapter focuses on the practicalities of information disclosure. It asks: how must this process be governed? A definition of governance as the process of steering a society toward a set of predefined goals is introduced. It discusses the benefits and difficulties of creating collaborative governance in the context of our project. The chapter presents our findings regarding governance from the I-Choose project. It evaluates existing experiments in collaborative governance that aim to extract public value from data disclosure, drawing on several examples from multiple countries, including the I-Choose project. We find that information disclosure alone is not enough to enhance the public sphere. It must be supported by innovative governance mechanisms that address classic problems such as establishing independence among producing and regulating organizations and creating procedural transparency.

Keywords I-Choose • Data disclosure • Consumer choice • Governance

---

H. Jarman (✉)
Department of Health Management and Policy, School of Public Health,
University of Michigan, Ann Arbor, MI, USA
e-mail: hjarman@umich.edu

L.F. Luna-Reyes
Department of Informatics, University at Albany, Albany, NY, USA
e-mail: lluna-reyes@albany.edu

T.A. Pardo
Center for Technology in Government, University at Albany, Albany, NY, USA
e-mail: tpardo@ctg.albany.edu

H. Jarman, L.F. Luna-Reyes (eds.), *Private Data and Public Value*,
Public Administration and Information Technology 26,
DOI 10.1007/978-3-319-27823-0_8

8.1 Introduction

Technology is facilitating a revolution in the way we access information about markets, dramatically lowering the opportunity cost of learning about the provenance of the products and service we purchase and consume. As discussed in previous chapters, governments wish to encourage the disclosure of product data for a number of reasons, including facilitation of consumer choice, product innovation and research, and more efficient ways to regulate markets. This goal cannot be achieved without collaboration among a range of disparate actors in the supply chains, as well as others in the certification industry. Collaboration between government agencies, private actors, and consumer advocates is necessary in order to promote the disclosure of product data and to move toward the long-term goal of private sector transparency.

This book draws lessons from existing research for policymakers and other actors seeking to increase "private" data sharing to promote more effective and efficient public policies. As we established in the introductory chapter, an enormous amount of data about the provenance and safety of the products we buy, our healthcare, education, financial transactions, and many other activities regulated by governments, are held by private organizations. Much of this data is fragmented and separated from publicly assembled datasets. By creating incentives to share key elements of this data through a collaborative and consensual system, we can promote public value creation through reducing information asymmetries and providing better information for consumer's decision-making. As we argued in Chap. 2 of this book, it is also possible to find some applications that improve knowledge about some markets, which will be beneficial for supply chain participants. Additionally, the resulting combined and trusted data can be used for evidence-based policy decision-making, in market differentiation mechanisms, or to overcome market fragmentation in some sectors. Although this study focuses on one consumer product, coffee grown in Mexico and consumed in Canada and the United States, we believe that the lessons drawn are more broadly applicable to areas such as healthcare, trade, finance, food safety, and development.

This current and final chapter of the book takes one last pass at our initial question: how can we incentivize private actors to share their data in a way that promotes the public value of the information disclosed? As we have suggested throughout the book, answering this question does not only entail overcoming substantial technical challenges but also paying close attention to a series of individual, organizational, and institutional factors. Individual factors refer to the emergence of consumer demand for companies to disclose information about their products. This "demanded disclosure," driven by consumer demand for product data, is facilitated by new forms of technology that reduce the costs of exerting social pressure on organizations and governments (van der Laan, 2009). At the organizational level, issues of corporate social responsibility, private transparency, and opening government are the key drivers of these trends. Key institutions in the process, from our point of view, are those related to the certification and labeling mechanisms described in more detail in Chap. 4. Moreover, as it was demonstrated by the evaluation of our technical prototype in Chap. 5, even the ideal certification requires appropriate governance mechanisms to regulate decisions, standards, and general rules of the system. The following sections include reflections, explorations, and proposals on how a system of product data disclosure might be governed.

8.2 Key Governance Dilemmas for Data Disclosure

Although current and developing technologies make the disclosure and productive use of private and public data seem more feasible than ever before, data interoperability and disclosure are not solely problems of technology. They are also problems of human interaction. The competing interests of organizations and individuals with a stake in the debate over product data disclosure must somehow be managed. Furthermore, private sector organizations and public agencies have to be willing and able to share their data in a way that is useful and meaningful to consumers, stakeholder groups, and businesses.

There are several interconnected dilemmas facing anyone wishing to incentivize the disclosure of product data and promote its meaningful use (see Box 8.1). Several of these dilemmas have been already presented in more depth in one or more of the chapters in the book. The first set of dilemmas—discussed mainly in Chaps. 2 and 3—relate to the organization's decision to disclose data. As we know, any rate of data disclosure incurs some kind of cost. What is thus important for a private company to answer is whether the cost of disclosure is less than the anticipated benefits from disclosing the information and what is the necessary and desirable level of information that can be provided at a reasonable cost. The answers to these questions may be different for every organization, but they are also dependent upon what others decide to do. An organization's cost/benefit calculation might change, for example, based on the participation of a critical mass of similar actors or based on likely consumer demand for the disclosed data (Ran et al., 2016).

Policymakers must consider not just how to promote the disclosure of more information but how to improve the quality of the information disclosed—as discussed in Chaps. 5 and 6. We know that both government and private data often suffers from missing or incomplete information. The complexity of regulatory procedures means that there is considerable scope for errors and omissions. Government inspections of products might be patchy or inaccurately recorded. Compliance reports held by product certifiers may be submitted in hard copy only or in a format (such as PDF) that makes it hard to repurpose the information they contain. Missing, incomplete, or poorly trusted data is problematic and undermines the fundamental goals of smart disclosure, open government, or private sector transparency. The poorer the data, the higher the cost of utilizing it for other purposes. Private actors or consumers might be uninterested in product data that is not of high quality. Broker organizations looking to develop consumer tools may well pass on the opportunity to use certain datasets if they calculate that the up-front cost of making the information usable is too high.

Box 8.1: Interconnected Governance Challenges Around Smart Disclosure

1. Promoting disclosure of information among a mixture of public and private bodies
2. Promoting participation in efforts to create standardized procedures for disclosure and data management
3. Improving the quality and completeness of disclosed information
4. Making supply chain data, government records, and product ratings interoperable
5. Incentivizing the use of collaborative standards by supply chain actors and developers
6. Predicting and managing the effects of these collaborative standards on organizational behavior
7. Predicting and managing the effects of apps using these collaborative standards on individual behavior
8. Promoting trust in the system by protecting consumer privacy

In addition to creating conditions that might increase information disclosure among private actors and ensuring high quality of the disclosed data, policymakers must also think about how to make the disclosed data interoperable in order to promote its meaningful use. As illustrated in Chap. 5, where we describe our preliminary prototype, this is an enormous technical challenge, and the success of any attempt to solve it rests on the degree of cooperation among the main actors in the system. Interoperability requires extensive collaboration between organizations and individuals, which ultimately rests on establishing trusted relationships among them (see Chaps. 3 and 7).

Underpinning these dilemmas is the fact that policymakers have to make some significant predictions about how individual consumers will behave. Smart disclosure policies imply that better data disclosure will lead to an improved information environment for consumers and will impact consumer choice. Producers and retailers are interested in disclosing product data because of the potential to differentiate their products within crowded markets, making them more visible to consumers. Providing trusted information about the origins of a product to consumers can enhance a company's sustainable credentials against its competitors. But this only works if consumer behavior is truly altered by the disclosure of product data.

We do know that consumers' trust in the data provided plays an important role in whether or not they use a particular system (see Chap. 3 and also Luna-Reyes et al., 2013; Sayogo, Zhang, Liu, Picazo-Vela, & Luna-Reyes, 2014; Sayogo, Zhang, Pardo, et al., 2014). Consumers should be protected from fraudulent use of disclosed product data. Any governance system promoting product data disclosure should consider the relationship between collaborative standards for governing product data disclosure and hard law remedies against fraudulent use of product data, certifications, or labels.

An additional key challenge relating to consumer trust is how to protect consumer privacy in an open and accessible system. How should individual and commercial privacy be balanced with appropriate, and broadly applicable, access to information? In order for them to trust the system, consumers should have the right to expect that important personal information will be kept private and the right to be protected from organizations that want to use disclosed information for direct marketing and scams.

8.3 Governance Experiments That Can Inform Private Sector Transparency

Governance is distinct from government in that it refers to a dynamic process of governing that involves coordinated action by a mixture of government and nongovernmental actors to achieve a policy goal (Rosenau, 1995, 2000). Via various governance processes, political agendas are debated and determined, decisions are made and refined, and the resulting policies are implemented, enforced, and evaluated.

The concept of governance, as distinct from government, can be particularly useful in areas where lines of authority and accountability are unclear or rapidly changing. This might be because both public and private sources of funding are used to implement a policy, because the problem to be solved cannot be addressed without considering the boundaries between different political jurisdictions, or because technology is rapidly making existing policies obsolete. All three of these issues arise when considering how to make product data public. Such data is likely to be held by private organizations but regulated in the public interest, its creation and dissemination crosses the boundaries of multiple regional and national governments through global supply chains, and the means of gathering, transmitting, disclosing, and analyzing it are rapidly changing.

No two governance systems are the same, but forms of governance can be easily classified into three broad categories: hierarchical governance, market governance, and network governance (Powell, 1990; Thorelli, 1986). The contrasts between these three categories can assist us in evaluating how product information is currently governed and identifying potential improvements to that process.

Hierarchical governance is top-down and highly structured, with much of the authority to make and enforce policy decisions remaining with government (Börzel & Risse, 2010; Hill & Lynn, 2004). We might refer to this type of governance as "direct" regulation, where government agencies write rules and regulations based directly on legislation. Nongovernmental actors are consulted in this process, but the agency's decisions are ultimately binding upon them. Rules made through hierarchical governance are legally enforceable and often come with steep penalties for noncompliance.

In contrast to hierarchical governance, market governance is bottom-up. In market governance, governments rely heavily upon competition among private actors in order to incentivize desired behavior and achieve a policy goal (Donahue & Nye,

2002). Governments may rely on market actors to self-regulate, to introduce competition into sectors dominated by publicly funded service providers, offer incentives for certain actions through the taxation system at the organizational or individual level, or influence businesses through trade policy or investment guarantees.

Network governance can be thought of as more horizontal, although all networks contain some actors which are more powerful than others (Ansell & Gash, 2007; Goldsmith & Eggers, 2004). This form of governance relies on collaboration between peers or peer organizations, both to make policy decisions, and to change the behavior of participants. The relevant network might be brought together around a common goal or set of norms or drawn from an economic sector, industry, or profession.

The three models of governance outlined above are, of course, ideal types. Most governance systems contain more than one of these elements, and some contain all of them. They are, however, useful analytic tools. Over time, the trend has been for governance systems in many policy areas to become less hierarchical and more reliant upon market mechanisms to incentivize certain desired behavior. Governance processes have in many cases become more transparent, with consultations between government and nongovernmental organizations more visible and accessible to less powerful groups. In some cases, particularly in cross-border regions such as the European Union and ASEAN, experiments in network governance have multiplied in the last decade (see Box 8.2).

In the context of our project, it is inaccurate to speak of a single form of governance for international trade. A hierarchical global system—created by states, consisting of binding legal rules, and headed by a formal international institution, the World Trade Organization (WTO)—exists alongside a growing number of market and network-based regulatory and certification systems covering many products and services (Luna-Reyes, Andersen, Andersen, Derrick, & Jarman, 2012). There is a rapidly growing literature on these systems, which include standards produced by the International Organization for Standardization (ISO); on products produced by transnational private sector regulators such as financial product rating agencies, industry-wide accreditation bodies, international commercial tribunals, and nongovernmental product certification schemes (Auld, Cashore, Balboa, Bozzi, & Renckens, 2010; Bartley, 2003; Büthe, 2010; and on voluntary programs Coglianese & Nash, 2001; Darnall, Potoski, & Prakash, 2010).

There is an important distinction between how state and non-state regulatory systems treat non-price information. The state-led system intentionally separates economic issues from the social and environmental consequences of trade, with the aim of identifying and eliminating trade barriers (Jarman, 2009; Jarman et al., 2011). The very purpose of non-state systems, however, is often to combine price and non-price information in order to increase the desirability and the value of the goods being traded (on coffee, e.g., see Fridell, 2007). At the moment, these two types of systems are not particularly compatible. As FIPP usage grows, however, it may become increasingly difficult, or less desirable, to separate price and non-price information, challenging the existing state-led trading system and presenting governments with a range of new regulatory choices.

Box 8.2: The EU's Platform for Diet, Nutrition, and Physical Activity

The EU's Platform on Diet, Nutrition and Physical Activity is an example of hybrid governance. At the time of the Platform's launch in March 2005, the European Union's central organizations had little independent authority to make public health policies affecting diet and exercise (Jarman, 2009). The European Commission, one of the EU's main agenda setting bodies, sought to build a coalition of European governments, private retailers and the food industry, public health advocacy groups, and consumer NGOs in order to "pool expertise and best practice" and act as a "catalyst for action" on diet, nutrition, and physical activity (ICF Consulting, 2015). Organizations and governments were asked to make public pledges regarding diet and exercise policy: fast-food companies might pledge to reduce fat and salt content, for example, while nonprofit groups could pledge to increase information campaigns regarding the benefits of physical activity, or local governments could implement plans for increasing access to bike paths.

The Platform relied heavily on peer pressure and data disclosure to hold organizations and governments to the pledges made, including independent reviews of progress that were made public (Hallsworth & Ling, 2007). Participants made commitments on a range of activities as diverse as improving food labeling, the reformulation of food products, modifying portion sizes, limiting food advertising to children, and implementing lifestyle programs in the workplace (Hallsworth & Ling, 2007).

The pledges made were not legally binding, and monitoring progress through third-party reviewers was not intended to be a "value judgment on the activity itself" but rather a means whereby the completeness and timeliness of progress reports could be improved.

The success of this pledge and progress monitoring mechanism relied heavily on the "public visibility" of the Platform among nonparticipating organizations and the public (De la Porte, 2010). Policymakers anticipated that NGOs would be able to hold companies to their pledges by threatening to damage firms' reputations among consumers, but the Platform's poor public visibility meant that the costs of noncompliance were low. Ultimately, many of the pledges were not fulfilled, and several headline participants pulled out of the program prematurely. The Platform, however, built enough of a consensus toward the establishment of new EU-wide powers and policies in public health.

The governance of product information reflects these overall trends in governance models. Private actors currently hold the majority of information about how products are grown, manufactured, distributed, and sold. Consumers can access only observable information about the products that they buy, traditionally relying heavily on producers, distributors, and retailers, and associated marketing and brands, to provide them with information about their purchases. The relationship between producers and consumer demand is key to understanding producer decisions to disclose or not to disclose data about their products.

Nevertheless, governments play an important role in regulating various aspects of both products and supply chains. Government regulations prohibit or limit the inclusion of certain chemicals or ingredients for public health reasons and put limits on polluting emissions from the manufacturing process. Government agencies regulate markets, brands, advertising, and financial transactions, and products moving across borders are inspected by customs agencies. Governments also set rules regulating the information that companies provide to consumers about their products, mandating some information and overseeing redress procedures addressing misinformation. International and supranational organizations (e.g., the WTO or the European Union) and the agreements which underpin them (the North American Free Trade Agreement) further regulate activities such as trade and other forms of economic exchange and promote the coordination and harmonization of domestic regulations.

Importantly, nongovernmental regulation of product information has flourished in recent years as we discussed in more detail in Chap. 4. Products claiming to have distinct non-price characteristics (such as organic status) are certified by third-party organizations and government agencies along specified criteria. Certifying organizations can be regional, national, or international, with the smallest often organized into larger networks. The implications of this explosion of interest in product certification and labeling for smart disclosure and private sector transparency are significant.

8.4 The Strengths and Weaknesses of Certification and Labeling

One of the key assumptions of smart disclosure and private sector transparency policies is that information can be used as a policy tool to manipulate markets (see Chap. 2 and Sect. 8.2 above). As we further discussed in Chap. 4, there are two main ways of using information to intervene in markets: direct and indirect intervention (Weiss, 2002). Direct intervention occurs when a government collects and distributes information directly to the public. Indirect interventions occur when other actors generate or share the information that is required or enabled by the government. Product labeling is one example of an indirect intervention.

Indirect interventions such as product labeling schemes can either be mandated by governments or can remain voluntary. Mandatory policies or standards require the compliance of all market actors without any exception. These policies are supported by government regulation through a process of hierarchical governance (see Sect. 8.3, above). For example, the US government required all commodity product manufacturers to attach nutrition labels to their products based on the Nutritional Labeling and Educational Act of 1990.[1] Another example is the Alcohol Beverage

[1] http://www.fda.gov/ICECI/Inspections/InspectionGuides/ucm074948.htm

Labeling Act of 1988, which requires two mandatory warnings to be placed on alcoholic beverage containers.

Voluntary policies allow market actors to adopt or ignore measures as they see fit. Most labeling and certification regimes are voluntary procedures whereby organizations complying with a certain standard or set of standards may attach a label into their product. Organizations use such labels to signify their credibility to consumers regarding claims they make on certain issues such as environmental sustainability or human rights.

Voluntary policies exist under two different types of governance regime. Under hybrid systems of governance, voluntary standards can be created or enabled by government but administered by an independent body. For example, the government of Quebec legislatively authorized an independent organization—CARTV—to monitor labeling systems regarding the origin and authenticity of products sold within the province.

Under market governance regimes, standards are "buyer driven." This means that their adoption among organizations, and the rate of compliance, depends on market demand for the standards (Cashore, Auld, & Newsom, 2004). These kinds of standards tend to proliferate within markets due to asymmetric information between consumers and producers (Jahn, Schramm, & Spiller, 2005).

As we discussed in Chap. 2, under conditions of asymmetric information, consumers have limited ability to assess product attributes, especially credence and/or quasi-credence attributes. Credence attributes (such as the claim that a brand of toothpaste provides plaque reduction) are difficult for the consumer to assess, even after using the product. However, they can be verifiable through inspection by external parties (Albersmeier, Schulze, Jahn, & Spiller, 2009). Plaque reduction, for example, could be verified through scientific testing. Quasi-credence or "Potemkin attributes" (Jahn et al., 2005) are similarly difficult for the consumer to evaluate but are related to internal production methods and processes (Albersmeier et al., 2009). Methods for evaluating quasi-credence claims are still being explored. A claim that a brand of organic coffee beans protects the environment more than a nonorganic brand is a quasi-credence, or Potemkin attribute. This claim can be evaluated by a third-party organization via a process of certification, with this process reflected in the product labeling.

As an alternative to relying on brand recognition alone, the main objective of voluntary standards is therefore to provide assurance to consumers regarding the credibility of quasi-credence claims represented by product labels. The ISO 14020 standard, an internationally recognized information standard established via a process of expert deliberation, classifies certifications and labeling schemes based on the extent and quality of information verifiability in the label. The ISO 14020 divides the certifications scheme into[2]:

[2] For further comparison among the three types, refer to http://ec.europa.eu/environment/ecolabel/about_ecolabel/reports/erm.pdf

Type I labeling scheme (ISO 14024): a labeling scheme that provides a seal of approval based on the verification of attributes by independent third-party certifiers or verifiers. An example of this type I labeling schemes are the third-party certifications such as Fairtrade, UTZ, Rainforest Alliance, and others.

Type II labeling scheme (ISO 14021, 14022): a labeling scheme that is self-declared by private organization without a verification from independent third-party certification. An example of this type of labeling is the C.A.F.E. Practice by Starbucks.

Type III labeling scheme (ISO 14025): a labeling scheme that targets more toward business-to-business relationships. The claims for this type of labeling consist of quantified product information based on product life-cycle assessment (LCA). This labeling scheme aims to provide common label parameters and methodology that would enable the comparability between products.

The rapid growth in the number of different types of labeling schemes and third-party certifications have led researchers to question the utility of labels in assisting consumer choice in terms of both their credibility and quality (Raynolds, Murray, & Heller, 2007). The utility of these labels can be examined from three perspectives: the coverage or scope of the label, the governance of the certification scheme, and the assessment processes used when deciding whether or not to give permission to use the label. In Chap. 4, we presented a detailed review of current labels in the market and their assessment. Our review concluded that certifications emphasize transparency, legitimacy, and accountability of their practices. Majority of the certification and labeling schemes openly publicize their standards and principles to demonstrate the transparency of their governance process. They also use a number of approaches to assert the legitimacy of their practice, such as accreditation from reputable national or international organization. They put emphasis on the democratic nature of their governance processes, such as strong collaboration with NGOs and producers during the standard setting phase. Unfortunately, information about the steps certifications are taking to increase their trustworthiness is not readily apparent to the consumer. The information is unavailable either because consumers need to expand significant effort to research this information or some of this information might be proprietary. In addition, even if this information was available to the consumers, the magnitude of this information might deter consumers from using it as a basis for their purchasing decisions.

As we have proposed in Chaps. 5 and 6 of the book, one way to address these issues is to build an interoperable data platform that would enable private sector actors to share information and data through the use of agreed-upon semantic and ontology standards. Such platform can enable data owners and producers to make their information readily available by standardizing and simplifying the process for data publishing. The platform can thus promote the creation of an ecology of producers, consumers, and certifiers that produce, prepare, and consume information that is designed to empower a consumer to make a purchasing decision consistent with their values.

Development of a system based on an interoperable data platform designed to support private sector transparency should follow these principles:

1. *Human collaboration as the foundation of a data commons.* The first goal of any data sharing initiative should be to create good working relationships between its participants. In bringing actors from very different fields together, collaborations should aim to build consensus as to the goals and direction of the project, focusing on the people involved as well as technological solutions. Chap. 7 in the book suggests to pay attention to key values of participants (industrial, civic, market and domestic) to be able to reach such agreements.
2. *Creation of transparent and open data systems.* Participants should share the goal of increasing data transparency through an open system that is collectively owned and managed.
3. *Maintenance of brands and promote market differentiation.* Any collaboration should be flexible enough to enable private commercial actors to maintain the values of their brands, and ideally, allow these actors to improve their differentiation in the market by enabling them to promote the products of the data commons to their customers.
4. *Ensurance of consumer protection and privacy.* Any data collaboration should protect the rights of consumers to keep their personal information private. Consumers should be protected from organizations which might engage in marketing scams.
5. *Support of equal access to decision-making.* Any collaboration should strive from the outset to include representation from groups and individuals who have fewer opportunities to access the shared information and to give these groups equal decision-making power. These groups might include consumer or patient representatives, new or small providers or producers, or grassroots and community groups.

Using these five principles, we can outline five actions to encourage disclosure of private data. The following sections will describe these actions in more detail:

1. *Creation of a data commons.* A group of interested organizations and individuals collaborate to create a "data commons"—a set of common information that is shared among the group in order to achieve collectively agreed goals. A subset of this information is released to the wider public via an API. One of the key tasks for the group is to collectively agree on the scope of the data commons and the extent of data to be publicly released.
2. *Progressive release model.* Although the initial amount of data in the commons may be small, participants commit to progressively increase the amount of data that they release, setting goals and benchmarks for this process which are approved by the group and made public.
3. *Peer review of process, supported by* "digital proof" *lemon laws.* A peer-reviewed process among the participants monitors the amount and quality of the data released and puts pressure on organizations which fall behind on releasing data against their stated benchmarks. To protect against falsified information, a claim by a member against one of the data providers triggers a review. If the review process determines the information is false, there is a legal duty for the organization to correct the data. Existing regulations are reviewed to gauge their ability to support this system. New regulations are created if necessary.

4. *Rewriting of organization DNA*. Participating organizations rewrite their founding documents (e.g., articles of association) to reflect the group's consensus regarding the goals and key principles of data sharing.
5. *Balancing of power structures*. The group's decision-making structure promotes the representation and inclusion of weaker actors in group decision-making. Efforts are made to minimize the resource burden of participation. Members make formal public statements about their relationships with, and representation of, outside groups. The group publishes and promotes model partnership agreements which others can use.

8.5 Collaborating to Create a Data Commons

Chapters 3 and 5 of this book describe a platform enabled by the I-Choose architecture. The creation of such a platform, as suggested by the research presented in this book, requires at least three components: (1) a series of standards to share data in an interoperable and usable way (I-Choose ontology), (2) a series of standard application programing interfaces (APIs) to allow developers to create applications and take advantage of the information being shared, and (3) a collaborative governance system involving consumers, government, NGOs, industry associations, etc. to facilitate the creation and maintenance of both sets of standards. Our conceptual data sharing architecture is shown in Fig. 8.1.

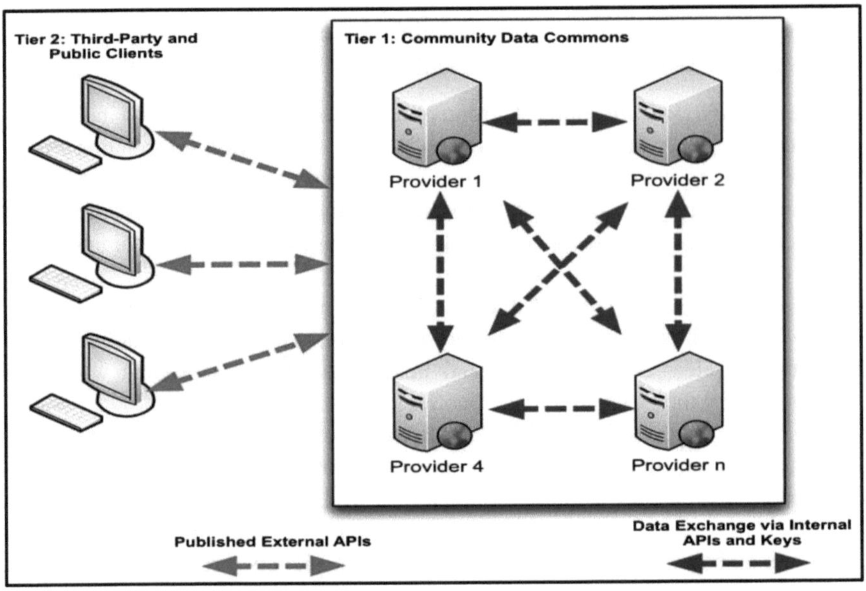

Fig. 8.1 Two tiers of data sharing (Luna-Reyes et al., 2012)

In the data commons, the participating members will agree to an interface and data specifications for a web service that each member must implement to be considered part of the community. The standard for data exchange will be an encapsulated set of internal (i.e., community) web APIs. These community specifications will be a defined set of Hypertext Transfer Protocol (HTTP) request messages coupled with the definition of the structure of the expected response messages. The data common's standard will specify data responses that are complete, consistent, and accurate and will evolve over time based on the community's desires. The standard will also specify the standardized web port for data exchange and the exact format of the requests and responses (e.g., Extensible Markup Language (XML), JavaScript Object Notation (JSON)).

Each participant in the data commons community will be responsible for implementing the agreed-upon standard by developing and deploying the web service specification. This will allow each organization the ability to maintain their own database systems but facilitate the free flow of information between the community members. The producers will issue data keys to other members of the community, which will control access to the data commons. Each producer will be able to ensure the privacy and integrity of their data, and members of the community can self-regulate and verify other members by simply sending the agreed-upon data requests to members' web services and then examining the returned responses to ensure compliance with the standard.

A similar process will be applied to provide access to third parties and the public. A "non-keyed" data request format will be published as the standard, which will allow the release of a subset of data to noncommunity members. The advantage of using these web service APIs is that it will allow for development of "mash-ups" using the public data from the community for easy comparison and consumption. A mash-up is an application or web page that combines and uses data from multiple sources. Mash-ups can provide fast and easy integration using the published community APIs and data sources from the various producers. This architecture will also allow mobile device developers to create mobile applications that leverage all of these new sources of information.

Of course, a reliable and trustworthy system requires, as it was suggested in Chap. 2, peer-reviewed mechanisms to encourage the disclosure of product data. In the following paragraphs, we outline our strategies for peer review within the data commons and the progressive release of supply chain data through an iterative process.

Agreeing on the necessary data standards need not involve imposition. Handling this problem requires delicate negotiation and collaboration within a framework that engenders trust between organizations, as well as between consumers. Any governance system, which aims to promote FIPP regimes, will need to bear different configurations, participants and values in the supply chain as we discussed in Chap. 7. Conflicting values between supply chain participants may prevent any cooperation. Overzealous regulation may turn off producers and retailers if the price is not right. At the same time, government-funding constraints critically shape regulatory capacity and direction. A key example is recent legislation[3] directing the FDA

[3] FDA Food Safety Modernization Act. 2011. H.R. 2751. http://thomas.loc.gov/cgi-bin/bdquery/z?d111:H.R.2751

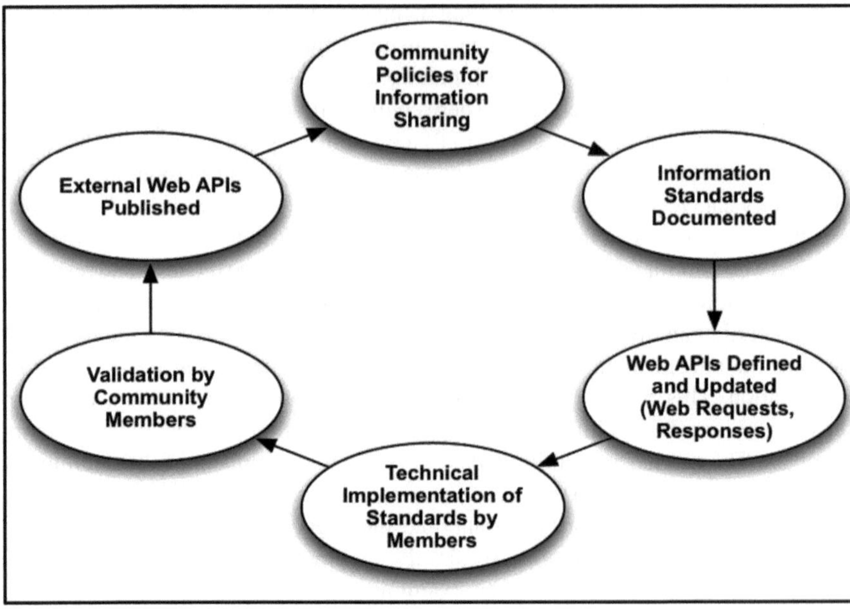

Fig. 8.2 Iterative process for policy and technical standard development and implementation

to improve product traceability, with the particular goal of improving food and drug safety. The FDA may not currently have the capacity or resources to make this mandate work. A danger exists that such systems will be seen as purely or mostly revenue raising mechanisms, charging large registration fees for entry, with little real oversight exercised. For this reason, as we explained in Chap. 2, focus group participants stressed the need for broad participation in, and ownership of, any governance structure.

Governments already have well-established standards for public and interest group consultation on trade, public health, and environmental issues—requirements to publish their actions and hold public meetings—as well as established advisory group systems that bring select stakeholders into regular contact with officials. Some of these public meetings, such as recent meetings to discuss the FSMA, already make extensive use of online tools. But great potential exists for governments to do more to bring consumers into the negotiating process—to make them part of the decision-making structure, not just an adjunct to it. For trade in goods, this means building a system that encourages broad consumer participation in formulating expectations for how corporations should act, incentives to support compliance with those expectations, and encouraging innovation among producers, suppliers, and entrepreneurs. We envision an iterative approach for developing and technically implementing the data sharing standards (see Fig. 8.2). The iterative development process together with the concept of a data commons can be also applied to other domains besides sustainable coffee supply chains. For instance, the federal government is currently incentivizing the creation of health information exchanges (HIEs) that would allow medical and public health data to be collected and made easily accessible to patients and healthcare professionals. One of the biggest challenges facing the creation of health information exchanges that are sustainable in the longer term is establishing collaborative relationships between fragmented private actors in the health sector. Creating a governance structure based on the principle of data commons could help with overcoming this barrier. The same could be said for an application in trade policy development, where such process could be used to streamline the production of a classification system which harmonizes product or service codes among the three countries of North America, which can currently take years to develop.

8.6 The Relationship Between Collaborative Standards and Hard Law

In order for any governance system to work effectively, participants must work together to achieve an acceptable balance between incentives for action and penalties for noncompliance. While hierarchical governance systems are defined in part by their clear lines of authority and binding regulations, most market and hybrid governance systems fall under the category of "soft law," meaning that while they create official documents such as guidelines, benchmarks, or collaborative standards, these commitments are not binding on the actors that make them—noncompliance cannot be judicially challenged.

Soft law is frequently used where boundaries between legal jurisdictions are messy or make problem solving difficult, such as in international contexts. UN resolutions and EU communications are examples of soft law. Not all international law is easily classified in this way—the NAFTA side agreements on labor rights and the environment are technically binding on North American governments but are largely voluntary in practice (Kirton & Trebilcock, 2004). Nevertheless, any prospective governance system for managing product information must take this balance between soft and hard law, between consensus and binding commitments, into account.

Importantly, soft law carries with it the expectation that it may one day be transposed into hard law. In other words, that the consensus built through collaboration, as expressed in the standards or guidelines created, will form a platform for future legislation. This is an important expectation, because there are things that soft law cannot do or does not do well. In particular, soft law does not protect consumers from organizations who are not just noncompliant but that are actively trying to do harm. This might include the fraudulent use of product labels, attempts to acquire personal information and abuse this information for profit, or other misleading product claims. Coffee producers in Mexico, as well as certifiers, for example, expressed their interest of having harder regulations regarding fair trade similarly to the clear rules of organic markets.

In order for any private sector transparency initiative to be credible and widely used, the governance system created must be able to relate soft law collaborative standards to existing hard law. In the United States, for example, there are many governmental entities that regulate green marketing claims, including the National Advertising Division of the Better Business Bureau (NAD), Federal Trade Commission (FTC), Environmental Protection Agency (EPA), Department of Energy, National Association of Attorney Generals (NAAG), US Department of Agriculture (USDA), and the Food and Drug Administration (FDA). The FDA is the investigative/enforcement arm of the NAD. NAD "seeks to ensure that claims made in national advertising are truthful, accurate and not misleading. It requires that objective product performance claims made in advertising be supported by competent and reliable evidence" (Protecting consumers from false and deceptive advertising of weight-loss products, 2014). The NAD has issued nearly 30 decisions involving a wide range of "green" marketing claims, "often requiring the claims be modified or discontinued" as "each NAD case involves evaluation of the claims made in the advertising and labeling and the supporting evidence provided by the advertiser" (Peeler, 2008). Compliance with the NAD is voluntary, but they enjoy a high rate of compliance with their green marketing decisions.

Currently, green labeling is prosecuted under the false advertising laws, statutes, and regulations. The legal remedies for false advertising vary depending on statute and forum but may include injunction, damages/restitution, recall, corrective advertising, or fencing in (the FTC uses this remedy to inform industry what they think their rule should be). The most likely class actions are cases in which the consumer was misled by a specific green label and paid a premium for the alleged benefit that was fraudulent. The most straightforward claim by a plaintiff is to assert that the defendant's label has violated FDA regulations and marketed a product that could mislead a reasonable consumer.

The FTC Green Guides (Green Guides) are the FTC's administrative interpretation of the law as applied to environmental claims. The Green Guides are not independently legally enforceable under federal law—the enforcement provision is under Sect. 8.5 of the FTC Act, which prohibits "unfair and deceptive" trade practices. But many states have incorporated the Green Guides as law, including New York, who has adopted them as to "recycled," "recyclable," or "reusable" claims. The application of the false advertising legal framework to labels such as organic, fair trade, and Rainforest Alliance is similar despite the fact that only the term organic is specifically certified and regulated by the federal agency, the USDA.

8.7 Concluding Remarks: Encouraging Private Sector Transparency

We have argued that there are three key elements that play a significant role in increasing the likely success of a product data disclosure system: the extent to which organizations can and do participate in the system, the extent to which individual consumers and other interested nonparticipant organizations can access information disclosed via the system, and mechanisms which improve the extent to which the system and the data it produces can be trusted.

These three elements—participation, access, and trust—are interconnected. Participation relates closely to the "institutional visibility" of the system, the extent to which potential participants are aware of the system and its goals. Access relates closely to the "public visibility" of the system, the extent to which nonparticipant organizations and the public are aware of it. Taken together, these two factors are essential components of creating trust in governance experiments and the data that they produce—in creating a legitimate governance process and in creating legitimate policy outputs from that process.

Broad participation is key to the success of any attempt to increase product data disclosure or private sector transparency. By the term "broad participation," we are not just referring to having a critical mass of businesses in a particular sector or industry as participants but to the range and type of actors which are active participants in the system. A broader range of actors improves the potential for innovation. It also prevents regulatory capture.

SMEs, small developers, and other groups may lack expertise and need support to participate in the scheme, while producers and suppliers with fewer resources may have limited access to the Internet or technology. Current use of information systems and technologies varies in an important way not only across market actors but also among other potential members of collaborative networks such as consumer groups or other NGOs. For instance, coffee cooperatives traceability systems of organic and fair trade coffee are very carefully designed paper-based system, while retailers like Walmart are heavily investing in advanced hardware and sophisticated information systems.

Scholarship analyzing interest group behavior shows that organizations with limited resources use networking and coalition building strategies in order to counteract their resource deficiencies to some degree. Advocacy networks and social movements frequently include coalitions of likeminded organizations working toward common goals and sharing relevant information. Small business and professional associations replicate this strategy for smaller commercial actors and individuals with common professional norms.

Any governance mechanism encouraging private sector transparency should consider this dynamic. Producers and suppliers with fewer resources should be incentivized to act collectively in order to overcome these difficulties. The governance system might incorporate the promotion and facilitation of collective agreements between organizations with different levels of resources that are participating in the same data commons, which set out clear and equal relationships between the parties.

One further barrier to participation and access is the ability of consumers to access information in their own language. Any governance system should consider the need to promote crowd-sourced translation of various data. These language barriers include some important legal requirements in some cases. In Canada, for example, there are legal requirements to translate official information into both

French and English. Translation may impose additional cost burdens in jurisdictions with multiple language requirements.

In the long run, governance structures should promote North American partnership agreements between organizations, which lay out what the expectations are for participants. The aim here would be to promote model agreements to participating certifiers and supply chain operators, producers, and other participating groups that draw on innovative experiments in intellectual property law and practice. While it may or may not be desirable to require such agreements, participants in the NATC process could at least become model organizations, putting pressure on each other to address these ownership issues and sharing best practice. Given the current enormous resource cost to companies of acquiring and defending patents, any FIPP governance structure should embrace the potential inherent in these experiments. To be successful, this governance mechanism should respect several principles:

- A living set of process management standards for providing, handling, and using product data.
- Common minimum information standards and reporting requirements for sustainable goods.
- Aim to overcome information asymmetries through gradually increasing access to supply chain information.
- Achieved through collaboration between regulators/private sector/consumer champions, who publicly set benchmarks and note progress toward those benchmarks.
- Regular peer and consumer review of standards and progress toward them.
- Certification of companies who regularly meet the requirements during peer and consumer review.
- Drafting and promotion of model "partnership agreements," which govern the relationships between strong and weak links of the supply chain (i.e., make sure that the producers don't get screwed).
- Data commons as a means to promote innovative, more open approaches to information ownership and intellectual property—evidence is that competition between companies gives way eventually to industry standardization.

One way to push this vision forward might be to establish a North American Traceability Council (NATC) with representation from stakeholders. Its members should include—at the least—government representation, sustainability and green supply chain experts, industry associations, and consumer organizations, with mandatory representation from or on behalf of producers and consumers with fewer resources. The NATC should be as "virtual" as possible to minimize costs to participants. The Council would work together to set the binding minimum information standards that will allow the system to function, based on existing information standards for traceability in the three countries.

The NATC could then oversee the negotiation and implementation of a "virtual ISO" which is not just negotiated in private by interested parties but is regularly reviewed by NATC and "consumer champions," active consumers who participate regularly in reviewing products and organizations. The ISO would become a living set of process management standards for providing, handling, and using product data. They would be regularly updated with input from stakeholders, consumers, and regulators. Companies, NGOs, or government bodies could be certified under this ISO if they consistently meet the requirements for peer and consumer review. Evidence from scholars of global policy suggests that ISO standards, such as ISO 9000, are popular with companies as means to differentiate themselves in the marketplace and seen as both flexible and fairly objective.

Within the framework of the virtual ISO, companies could nominate key substantive parameters and set benchmarks against them. Participating companies would be free to use the aggregate information in their product marketing, creating their own labels and so on, but under advisement from the NATC and guidelines established through the ISO. Results of peer review would become incorporated into future versions of the ISO. This is not a pipe dream—elements of this system are currently being implemented in several other countries.

This peer-reviewed process would be matched by consumer review. Information from participating consumers acting in social networks—such as product and company reviews or location information enabling better supply chain mapping—could contribute to the evidence base for future versions of the ISO and for benchmarking exercises. Consumers could opt to comment on and review products, producers, and other suppliers. They would benefit by seeing what others have recommended and use this to assist their purchasing choices. Companies, on the other hand, would gain the opportunity to "test" the sustainability of their products among participating consumers and create better connections. NGOs could submit information into the system about certification, inspection, and production practices. Reviews and reports would be verified and ranked by other participants. Consumers who submitted highly ranked product reviews consistently would become "consumer champions." Results from consumer comments could be aggregated and accessed by all participants. Significant consumer concerns would be addressed through the peer-reviewed process.

References

Albersmeier, F., Schulze, H., Jahn, G., & Spiller, A. (2009). The reliability of third-party certification in the food chain: From checklists to risk-oriented auditing. *Food Control, 20*(10), 927–935. http://doi.org/10.1016/j.foodcont.2009.01.010.

Ansell, C., & Gash, A. (2007). Collaborative governance in theory and practice. *Journal of Public Administration Research and Theory, 18*(4), 543–571. http://doi.org/10.1093/jopart/mum032.

Auld, G., Cashore, B., Balboa, C., Bozzi, L., & Renckens, S. (2010). Can technological innovations improve private regulation in the global economy. *Business and Politics, 12*(3), 1–42.

Bartley, T. (2003). Certifying forests and factories: States, social movements, and the rise of private regulation in the apparel and forest products fields. *Politics & Society, 31*(3), 433–464.

Börzel, T. A., & Risse, T. (2010). Governance without a state: Can it work?: Governance without a state. *Regulation & Governance, 4*(2), 113–134. http://doi.org/10.1111/j.1748-5991.2010.01076.x.

Büthe, T. (2010). Private regulation in the global economy: A (p)review. *Business and Politics, 12*(3). http://doi.org/10.2202/1469-3569.1328.

Cashore, B. W., Auld, G., & Newsom, D. (2004). *Governing through markets: Forest certification and the emergence of non-state authority*. New Haven, CT: Yale University Press.

Coglianese, C., & Nash, J. (Eds.). (2001). *Regulating from the inside: Can environmental management systems achieve policy goals?* Washington, DC: Resources for the Future.

Darnall, N., Potoski, M., & Prakash, A. (2010). Sponsorship matters: Assessing business participation in government- and industry-sponsored voluntary environmental programs. *Journal of Public Administration Research and Theory, 20*(2), 283–307. http://doi.org/10.1093/jopart/mup014.

De la Porte, C. (2010). State of the art. Overview of concepts, indicators and methodologies used for analyzing the social OMC. *Overview of concepts, indicators and methodologies used for analyzing the social OMC (November 1, 2010)*. REC-WP working paper on the reconciliation of work and welfare in Europe, (15/10). Retrieved from http://papers.ssrn.com/sol3/papers.cfm?abstract_id=1711245

Donahue, J. D., & Nye, J. S. (2002). *Market-based governance: supply side, demand side, upside, and downside*. Cambridge, MA/Washington, DC: Brookings Institution Press. Retrieved from http://www.brookings.edu/research/books/2002/market-based-governance

Fridell, G. (2007). *Fair trade coffee: The prospects and pitfalls of market-driven social justice* (Repr.). Toronto: University of Toronto Press.

Goldsmith, S., & Eggers, W. D. (2004). *Governing by network: the new shape of the public sector*. Washington, DC: Brookings Institution Press.

Hallsworth, M., & Ling, T. (2007). *The EU platform on diet, physical activity and health*. Retrieved from http://www.rand.org/pubs/technical_reports/TR474.html

Hill, C. J., & Lynn, L. E. (2004). Is hierarchical governance in decline? Evidence from empirical research. *Journal of Public Administration Research and Theory, 15*(2), 173–195. http://doi.org/10.1093/jopart/mui011.

ICF Consulting Services. (2015). *EU plataform on diet, physical activity and health – 2014* Annual report.

Jahn, G., Schramm, M., & Spiller, A. (2005). The reliability of certification: Quality labels as a consumer policy tool. *Journal of Consumer Policy, 28*(1), 53–73. http://doi.org/10.1007/s10603-004-7298-6.

Jarman, H. (2009). *Imagined commodities: "Trade and" policies in the European Union and United States*. London School of Economics and Political Science, UK. Retrieved from http://etheses.lse.ac.uk/2330/

Jarman, H., Luna-Reyes, L. F., Zhang, J., Whitmore, A., Picazo-Vela, S., Andersen, D. L., et al. (2011). *I-choose: Consumer choice, digital government, and sustainability in North America*. Presented at the APPAM Fall Research Conference, Washington, DC.

Kirton, J. J., & Trebilcock, M. J. (Eds.). (2004). *Hard choices and soft law: Voluntary Standards in Global Trade, Environment and Social Governance*. Toronto, Canada: Ashgate Publishing, Ltd. Retrieved from http://www.gowerpub.com/pdf/SamplePages/Hard_Choices_Soft_Law_Intro.pdf

Luna-Reyes, L. F., Andersen, D. F., Andersen, D. L., Derrick, D., & Jarman, H. (2012). Full information product pricing regimes: Policy implications for US-Mexico sustainable commerce. In *Puentes Symposium*, Houston, TX. Citeseer. Retrieved from http://citeseerx.ist.psu.edu/viewdoc/download?doi=10.1.1.310.4882&rep=rep1&type=pdf

Luna-Reyes, L. F., Zhang, J., Roy, R., Andersen, D. F., Whitmore, A., & Andersen, D. L. (2013). Information strategies to support full information product pricing: The role of trust. *Information Polity, 18*(1), 75–91.

Peeler, L. (2008, May 19). *Green packaging workshop – Comment, project no. P084200*. Retrieved from https://www.ftc.gov/sites/default/files/documents/public_comments/2008/04/534743-00053.pdf

Powell, W. W. (1990). Neither market nor hierarchy: Network forms of organization. *Research in Organizational Behavior, 12*, 295–336.

Protecting consumers from false and deceptive advertising of weight-loss products, § senate committee on commerce, science and transportation. (2014).

Ran, W., Jarman, H., Luna-Reyes, L. F., Zhang, J., Andersen, D. L., Tayi, G. K., et al. (2016). Supply-chain transparency and governance systems: Market penetration of the I-choose system. In J. Zhang, L. F. Luna-Reyes, T. A. Pardo, & D. S. Sayogo (Eds.), *Information, models, and sustainability: Policy informatics in the age of big data and open government*. Basel, Switzerland: Springer.

Raynolds, L. T., Murray, D., & Heller, A. (2007). Regulating sustainability in the coffee sector: A comparative analysis of third-party environmental and social certification initiatives. *Agriculture and Human Values, 24*(2), 147–163.

Rosenau, J. (2000). The governance of fragmegration: Neither a world republic nor a global interstate system. *Studia Diplomatica, 53*(5), 15–40.

Rosenau, J. N. (1995). Governance in the twenty-first Century. *Global Governance, 1*(1), 13–43.

Sayogo, D. S., Zhang, J., Liu, H., Picazo-Vela, S., & Luna-Reyes, L. (2014). Examining trust as key drivers in smart disclosure for sustainable consumption: The case of I-choose. In *Proceedings of the 15th annual international conference on digital government research* (pp. 137–146). New York: ACM. http://doi.org/10.1145/2612733.2612747.

Sayogo, D. S., Zhang, J., Pardo, T. A., Tayi, G. K., Hrdinova, J., Andersen, D. F., et al. (2014). Going beyond open data: Challenges and motivations for smart disclosure in ethical consumption. *Journal of Theoretical and Applied Electronic Commerce Research, 9*(2), 1–16. http://doi.org/10.4067/S0718-18762014000200002.

Thorelli, H. B. (1986). Networks: Between markets and hierarchies. *Strategic Management Journal, 7*(1), 37–51. http://doi.org/10.1002/smj.4250070105.

van der Laan, S. (2009). The role of theory in explaining motivation for corporate social disclosures: Voluntary disclosures vs "solicited" disclosures. *Australasian Accounting, Business and Finance Journal*, *3*(4), 15–29. Retrieved from http://ro.uow.edu.au/aabfj/vol3/iss4/2

Weiss, J. A. (2002). Public information. In L. M. Salamon (Ed.), *The tools of government: A guide to the new governance* (pp. 217–254). New York: Oxford University Press.

Appendix A
Methodology

A.1. Introduction

The research that supports this paper emerges from the ongoing work of the North American Digital Government Working Group, a consortium of researchers exploring the impact of product labeling, data architectures, and government-sponsored information policies on the market share of organic, fair-trade, and eco-friendly products in the NAFTA region. In these alternative markets, price is often complemented with information, transmitted through trusting networks or certification labels that convey the conditions under which a product is produced and distributed.

The project started with a preliminary exploration in which we completed a number of case studies from the NAFTA region (Luna-Reyes et al., 2013), where producers created and sustained a FIPP network to deliver products to end consumers with a value-adding information package that allowed them to appeal to specific consumer preferences for green products while at the same time realizing a price premium. This case-based work reinforced the experimental work of Komiak and Benbasat (2006) and demonstrated the importance of trust in FIPP networks and in the adoption and use of recommendation agents by end consumers.

Our initial explorations made evident the interest in improving sustainability in the region as supported by important side agreements to the North American Free Trade Agreement (NAFTA) oriented to ensure that firms active in the NAFTA region observe fair labor practices and strive to minimize the environmental impacts of their activities—(1) the North American Agreement on Environmental Cooperation (NAAEC) and (2) the North American Agreement on Labor Cooperation (NAALC). The Commission for Environmental Cooperation and the Commission for Labor Cooperation are two trilateral organizations established within NAFTA to monitor and promote these agreements.

The purpose of the research project was then to explore a set of government-sponsored product labeling and information policies that may have the potential to

H. Jarman, L.F. Luna-Reyes (eds.), *Private Data and Public Value*,
Public Administration and Information Technology 26,
DOI 10.1007/978-3-319-27823-0

supplement a compliance-enforcement approach with a more market-based voluntary approach that relies on shifts in producer and consumer behavior to significantly expand the share of worker and environmentally friendly products traded within the NAFTA region. We proposed that the market share for such products can mirror the recent rapid expansion in "organic" food products that followed the development and implementation of organic food labeling and packaging standards led by federal and provincial governments in the United States, Canada, and Mexico.

We decided to start our exploration of FIPP networks with coffee because the NAFTA region is both an important market for coffee and an important coffee producer. In addition, there exists in the NAFTA region a variety of coffee production practices reflecting values such as fair wages and environmentally friendly or organic production. Finally, coffee is produced within a commodity-driven supply chain whose pricing dynamics have been well studied and hence is amenable to a simulation approach that can explore the features of alternative production strategies in a flexible "what if" approach.

As shown in Fig. A.1, the research program involved several interrelated activities: we conducted field work, developed simulation models, and created and evaluated an

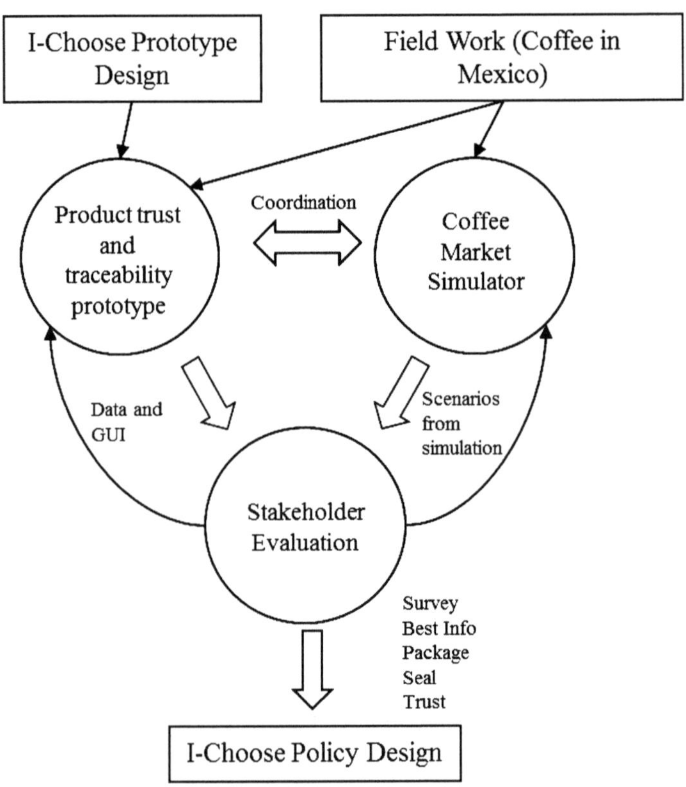

Fig. A.1 Overview of FIPP research program components and their relationships

information prototype to make policy recommendations. Although it was not part of the initial plan, we also conducted a survey to better understand how information packages in product labels were connected to different aspects of trust in the information included on it. In the following paragraphs, we briefly describe each of the components in this research program.

A.2. Fieldwork

A.2.1. Methods

The fieldwork component of our research project was intensive on interviewing. We interviewed multiple stakeholders along the coffee supply chain following a multiple case study approach (Yin, 1994), in which the analysis of individual cases is followed by an analysis aimed at comparing and contrasting their similarities and differences. To identify participants for the study, we followed a snowball sample approach, starting with interviewees from coffee associations who were then asked to identify other potential candidates. This process was reiterated with each consecutive interview. The semi-structured interview protocol included questions about information sharing, conventions on quality, characteristics of an information standard, motivations, barriers, and potential conflicts to share information across the supply chain. Each interview lasted approximately 1 h. The interview protocol was created originally in English by the project team. In order to be used in Mexico, the original protocol was translated into Spanish with the intervention of four project participants who were fluent in both English and Spanish.

We conducted 44 interviews in Mexico, the United States, Canada, and other countries such as Indonesia. Interviews took place in different moments during project development. We interviewed producers, intermediaries, roasters, and retailers. In addition, we interviewed some participants that were involved in more than one process, working, for example, as intermediaries and roasters, or more integrated operations that included production, intermediation, roasting, and final sales. We also interviewed members of coffee associations and certifiers. All interviews were recorded and transcribed with the exception of one where interviewee did not give permission for recording. In this case, three interviewers documented the interview. The chapters in this book use different subsets of these interviews in the analysis presented in each of them.

Interviews and case analysis helped us to develop a conceptualization of the main stakeholders and participants in the coffee supply chain. To compare and contrast responses from participants, we coded all the interviews. We worked together to define a set of 13 categories including themes such as barriers, motivations, values, conflicts, and qualities. We coded each interview according to these themes. In this way, by analyzing the interviews, we identified five different types of supply chains. We arranged codification tables belonging to each type of supply chain to understand the perspective of their members. Finally, we compared and contrasted the perspectives. Although we are reporting the results in English, data collection and analysis in Mexico was done in Spanish by native speakers.

A.2.2. Interview Protocol

This protocol is intended as a guide for the interviewer. After an introduction common to all interviews, it indicates key questions that will be asked. The overall structure of these questions will be consistent across all interviewees, but the more detailed follow-up questions will vary by interview to accommodate the interests of different kinds of interviewees (e.g., coffee producers versus regulators).

Date: ________________ Recording ref.: ________________
Time start: ____________ Location: ______________________
Time end: _____________ Interviewee: ___________________

Introduction

(a) Ask the interviewee to open the copy of project rationale/consent statement (already provided to them once by email or in hard copy prior to interview).

Interviewer: before we begin the interview, I need to go over some formalities. Law requires that research conducted by university researchers does not put participants at the risk of harm. This interview is completely voluntary and anonymous. You may choose not to participate, skip answering any particular question, and terminate the interview at any time. This interview will take approximately 1 h. All information collected will be kept confidential and you will not be identified by name in our report.

Before we begin the interview, I invite you to read the consent form (hand interviewee form). The purpose of this form is to inform you about your rights as a research participant. You can be assured that we will keep all information confidential and nothing you say will be attributed to you without your permission. [Allow time for reading.]

Do you have any questions regarding the interview or the form?

Interviewer: "I would like to start our discussion today by asking you some questions about yourself and your work. Your name or any information that might identify you as an individual will not be used in the final report. I would also like to ask your permission to tape this interview. As I indicated in my request for this interview and as you can see from the project statement, we are researching [project goals from this interview]. The taped records are only to support my notes and recollection of what we talk about today and will not be made available in their entirety to anyone outside the research team in taped or transcribed format."

Permission to tape: ______________________

"I would like permission to quote anonymously from the transcripts to support points made in the academic papers resulting from this project."

Permission to quote: ____________________

(b) Give interviewee an opportunity to ask questions of the interviewer, i.e., about the study, its goals, and the uses to which data will be put.

"Do you have any questions before we get started?"

____

Questions

1. *Introduction*

   I would like to take the first few minutes to hear about your organization and your role within it.

   (a) What are the major activities/mission of your organization?
   (b) What is your educational/professional background?
   (c) What is your official job title in this organization?
       And how long have you been in this position?
   (d) What are your roles/major responsibilities in this organization?

2. *How the I-Choose architecture will be used:*

   [We previously sent you two documents by email: (A) the mock-up application powered by I-Choose and list of competency questions and (B) the stakeholder map from I-Choose network meeting. Ask the interviewee to open the two documents].

   First, please open and look at document A, the mock-up of a future application we are envisioning. In fact, there are already several apps with similar functionality. This particular app, however, will go beyond what is currently possible because it will be powered by an interoperable information system called I-Choose. I-Choose can be understood as a huge virtual repository of data from the supply chain. We also have a preliminary list of questions in the second page that can be asked to this repository of data.

   Second, please open and look at document B, the stakeholder map that was built thanks to the inputs and comments from a group of participants in the coffee supply chain and other partners who met with us last August in Albany, NY. Now I would like to get your ideas to refine and expand these questions.

   (a) What questions might someone in your position ask of a system such as the one presented to you? What questions might your close partners' organizations ask of such a system? Why those questions?
   (b) What information is needed to answer these questions?

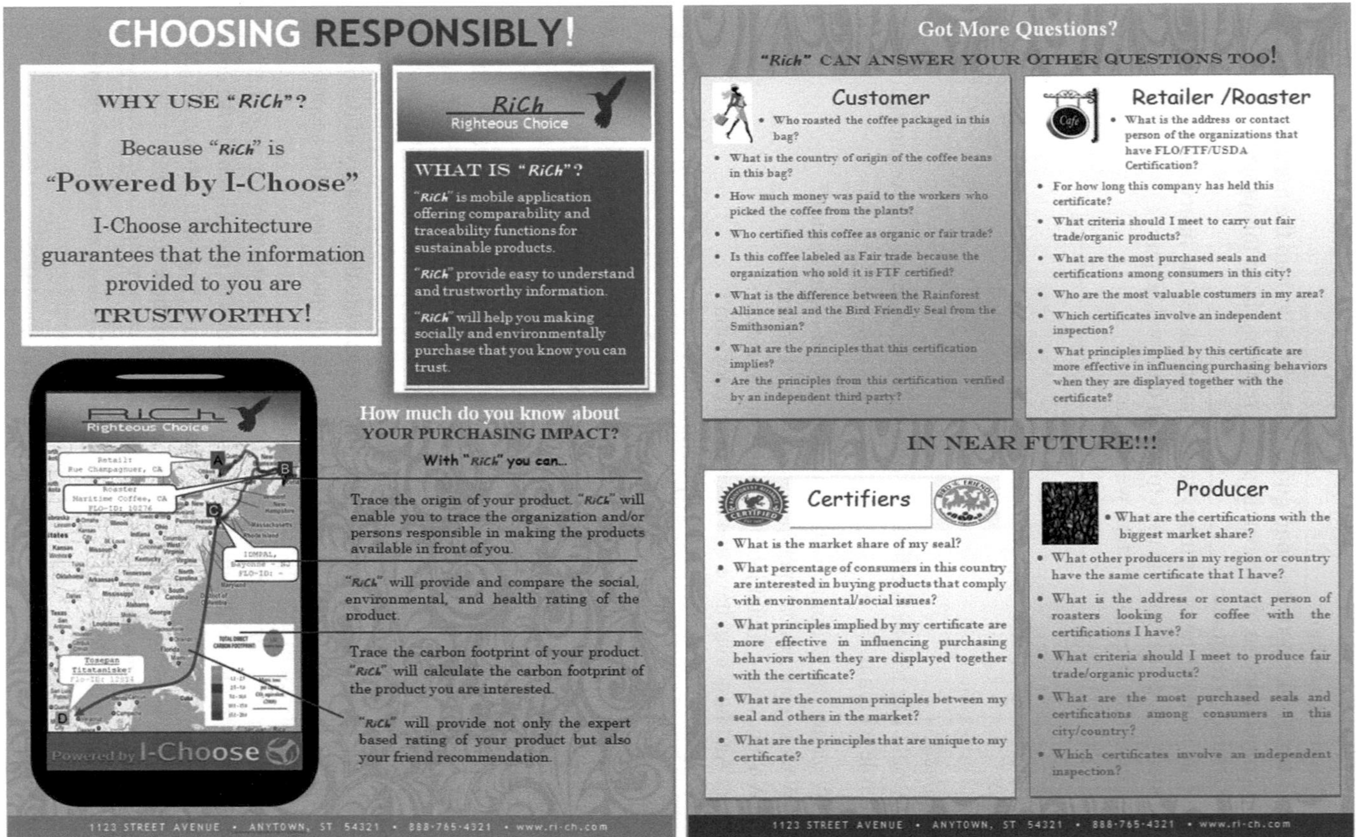

Fig. A.2 Attachment document A: mock-up application powered by I-choose and list of competency questions

Probes

(a) Are you aware of industry/government information standards to answer those questions?
(b) Are you aware of any current practices to use specific information to answer those questions?
(c) How do you currently get some of this information, if any?

(c) We know that new information systems can get subsequently used in unanticipated ways. Can you imagine any such uses of this system?

3. *How is information extracted from the supply chain?*

Now I would like to ask questions from the point of view of someone in your position and the point of view of your close partners.

(a) What do you see as the main motivators and incentives to share data to support a system like I-Choose from your perspective? What might your close partners' organizations see as the main motivators and incentives to share data to support such a system?

Probes

(a) Are these kinds of information available currently in your organization?
(b) What kind of information does your organization currently share?
(c) What level of detail of (proprietary) data or information does your organization allow to be shared?
(d) What level of detail (proprietary) data or information is your organization not willing to share?

(b) Systems such as I-Choose critically depend on the credibility and quality of available data. How can we create and maintain trust in a system like I-Choose?

Probes

(a) Who does what in terms of:

1. Policy
2. Management
3. Technology
4. Procedures
5. Laws and regulations

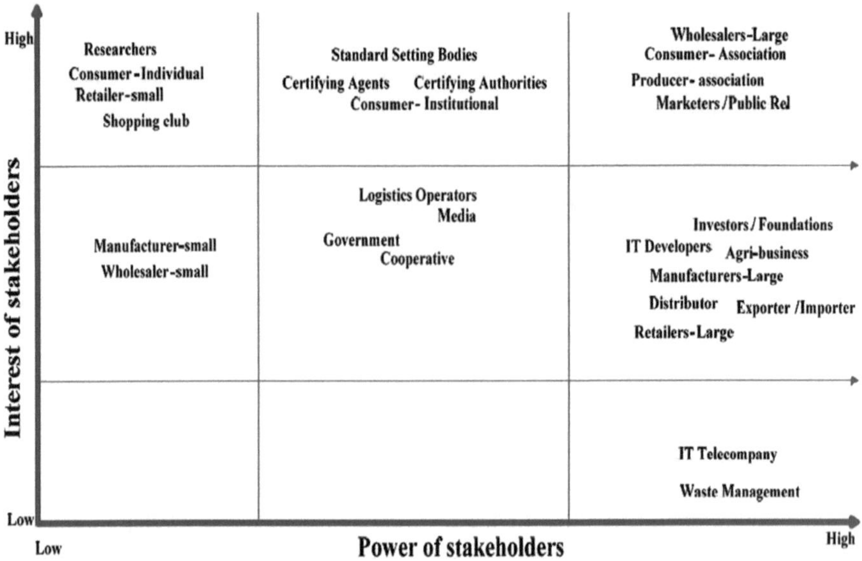

Fig. A.3 Attachment document B: stakeholder map from I-choose

(c) For your organizations, what are the main barriers to sharing trusted and quality data? For your close partners' organization(s), what are the main barriers to sharing trusted and quality data?

Probes

(a) In terms of:

1. Organizational processes
2. Strategies and goals
3. Legal requirements (security or privacy)
4. Industry regulations

4. *How can and should organizations like yours collaborate around a system like I-Choose?*

We expect that collaborating to share data within a system like I-Choose will depend upon building good relationships between organizations and people.

(a) What do you see as the main nontechnical or human barriers which might prevent organizations like your's from collaborating around a system like I-Choose?

Probes

(a) Which of these factors is most important?
(b) Are there ways in which these problems can be reduced or overcome? What are they?

(b) How do you envision the roles of the following organizations vis-à-vis a system like I-Choose in the longer term?

Probes
(a) Government
(b) NGO
(c) 3rd party certification bodies

(c) Are there other actors/organizations that should be involved in a system like I-Choose?
(d) Is a legal regulatory framework needed? If so, why is it needed? If not, why not?
(e) Do you see any problems in having different kinds of organizations participating in the same system (e.g., big and small organizations, commercial and nonprofit organizations, organizations from different countries)? If so, what are they? How might they be ameliorated or overcome?

(a) Can you think of any examples of existing systems where organizations collaborate to share data that might be good models for I-Choose? These could be government-regulated, commercial, nonprofit, or voluntary systems.

Probes
(a) In your opinion, why is that system a good model?
(b) To your knowledge, does/did that system experience any particular difficulties in its creation or management? Were these difficulties overcome? If so, how were they overcome?
(c) Is there something that organizations in an I-Choose system should do differently from the system you mention?

End

That concludes our interview. Is there anything else you would like to add or discuss in more detail? May we contact you again if we find that we need more information or clarifications?

Now that you are more familiar with our project, is there anyone else to whom you think would be interested in participating in our project? If so, could you provide them with our information?

Thank you very much for your time. We appreciate your participation.

A.3. Coffee Market Simulator

In this section of the appendix, we describe the processes involved in the development of a preliminary theory for governance and market penetration for the I-Choose platform. The process consisted of using system dynamics group model building as a tool to support interdisciplinary theory-building efforts. The modeling team included researchers involved in the I-Choose project who have been involved in the design and data collection processes. Overall, our research progressed through three methodological phases: (1) a large-scale concept elicitation meeting with stakeholders in the I-Choose supply chain, (2) a smaller-scale and more formal group model building project involving only team researchers who had been present at the larger stakeholder meetings, and (3) the creation of a simulation model.

A.3.1. Concept Elicitation with Stakeholders in the I-Choose Supply Chain

One main component of the project has been to create a network of researchers and key stakeholders of the coffee supply chain to understand the main requirements of a system like I-Choose. A core group of researchers from this network met regularly in a combination of face-to-face and electronic meetings from 2011 to 2014. This core group organized a workshop with a wider representation of stakeholders in a 2-day meeting in August 2011. The goals of the workshop were to understand which were the main stakeholders of a system like I-Choose and what were the key issues to be considered in the development of the system (Djoko Sigit Sayogo et al., 2015). The workshop involved a series of brainstorming and discussion sessions about main issues and stakeholders. Figure A.3 (above) shows one of the products of the meeting, in which participants identified and ordered key stakeholders according to their power and interest in I-Choose.

The theory-building process reported in this paper has been informed by this workshop and by a series of follow-up interviews with stakeholders that members of the core team have done during the last two years.

A.3.2. Group Model Building Exercise with Research Team

The second stage of this project component involved a series of formal and informal meetings to discuss and refine both model structure and behavior. Similar to many other group model building projects, we had a series of small scoping meetings with the small team of researchers working in the simulation model. As described in the literature, many different visual representations have been used during these meetings as boundary objects (Black & Andersen, 2012). These objects have helped this interdisciplinary team to communicate and work together sharing meanings and ideas. Figure A.4 shows what we understand as the root documentation of this project. The drawing represents I-Choose as an umbrella concept involving many

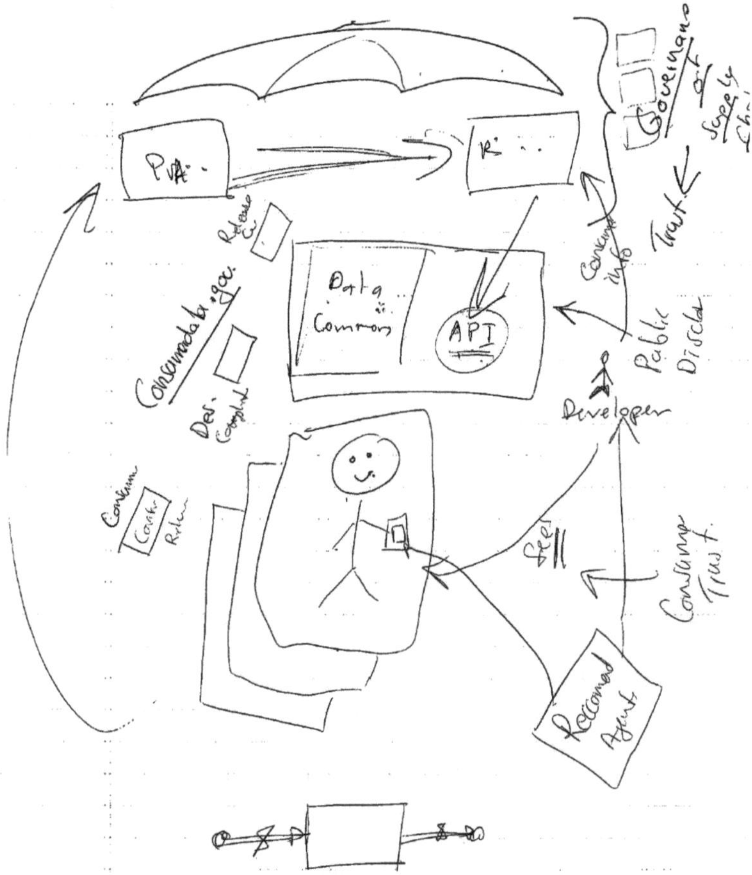

Fig. A.4 Preliminary concept drawing for the I-Choose simulation

components: supply chain participants joining an initiative, sharing data through a structure called the data commons, and building a set of APIs to make it possible for developers to create applications for consumers to use through mobile devices. Governance was considered in this preliminary drawing as a component that builds trust in the system.

Following these series of scoping meetings, we conducted a group model building exercise with members of the research team. The purpose of this stage was to create a dynamic theory of the growth of a market for "Green" product identification systems such as I-Choose. Each participant was asked to identify key variables in the I-Choose sociotechnical system and draw possible variable behaviors. Model variables were then selected based on participant votes and grouped into different clusters, and these clusters of variables and behaviors over time were used to build a sector view of the I-Choose system conceptual model. Figure A.5 shows a picture of the clusters of variables and behaviors over time created during the meeting. The center of the figure includes the sector view created from the clusters.

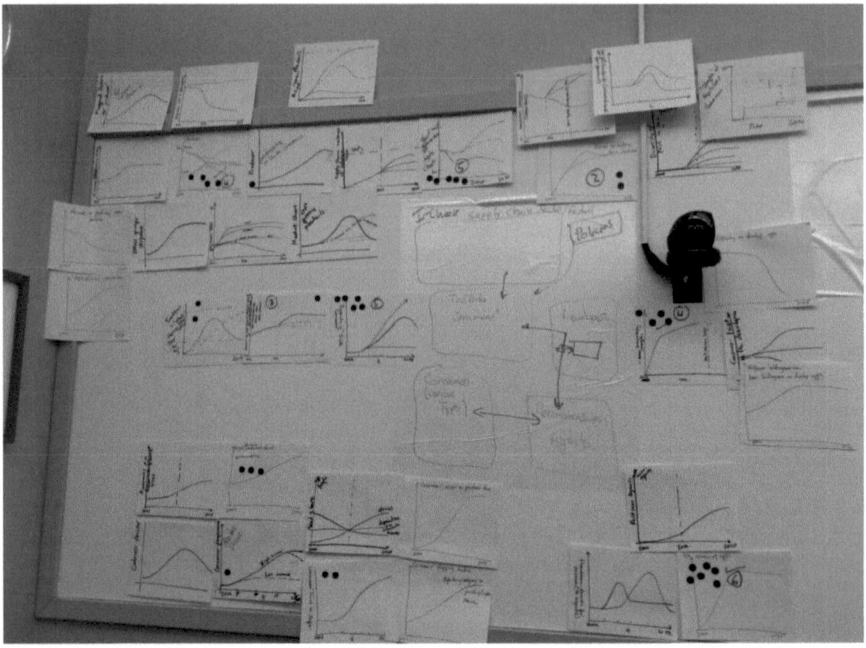

Fig. A.5 Model variables and a preliminary sector view from the GMB session

Based on selected key model variables, participants were asked to identify stocks and flows in the system and add causal relationships among these variables. The final product of this exercise is a preliminary conceptual model of the I-Choose sociotechnical system (see Fig. A.6).

Figure A.7 shows a clean version of some components of these two pictures. Figure A.7a has a clean version of the sector view in the center of Fig. A.5. The initial sector view considers also main causal relationships among these sectors. These main sectors were also reflected in the preliminary map from Fig. A.6, and A.7b shows a Vensim version of the picture in the board, including also the main areas showing main sectors in the model.

After eliciting the initial conceptual model—and recognizing the importance of the concept of governance—the group started a conversation around the concept, looking for important variables and concepts that could help to build an operational view of the concept. The group agreed on four important assumptions related to governance:

- Completeness: data should be complete and have a high quality.
- Openness: data should be accessible and transparent processes should be in place.
- Relevance: data should be relevant to the needs of consumers.
- Reliability: data should be accurate and the system reliable.

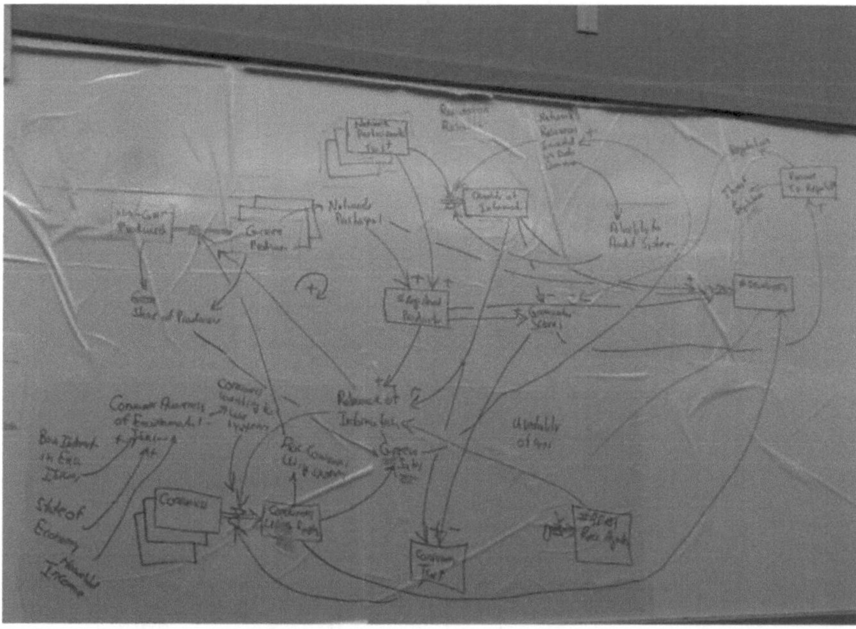

Fig. A.6 A preliminary conceptual model from the GMB session

A.3.3. *Model Formulation and Analysis*

With the products from the group model building exercise, we formalized the theory using mathematical formulations in Vensim. This process involved some additional thinking to make the conceptual thought operational, yielding a first running simulation model. We refined the preliminary conceptual model, identified key causal loops, and drew a system map, which provided guidance through the rest of the model building process. Figure A.8 shows one of these maps, with a more detailed and operational form of the Information Commons. As it is possible to see in the figure, some of the basic assumptions of governance were introduced into the conceptualization. These more operational conceptual models were refined and improved during a couple of scoping meetings.

While a running simulation model did emerge quickly from these sketches, its dynamics proved to be too complex or elusive for us to readily understand what was going on in the model. A persistent problem with these early attempts at simulating market growth and expansion was that the positive loops that we knew would dominate market growth seemed to be a too strong trap before the initial takeoff. We initially diagnosed this failure to take off as related to formulation problems in the Information Commons as shown in Fig. A.8. These initial formulations had failed to reveal what resources would be used to actually construct "system capabilities and processes".

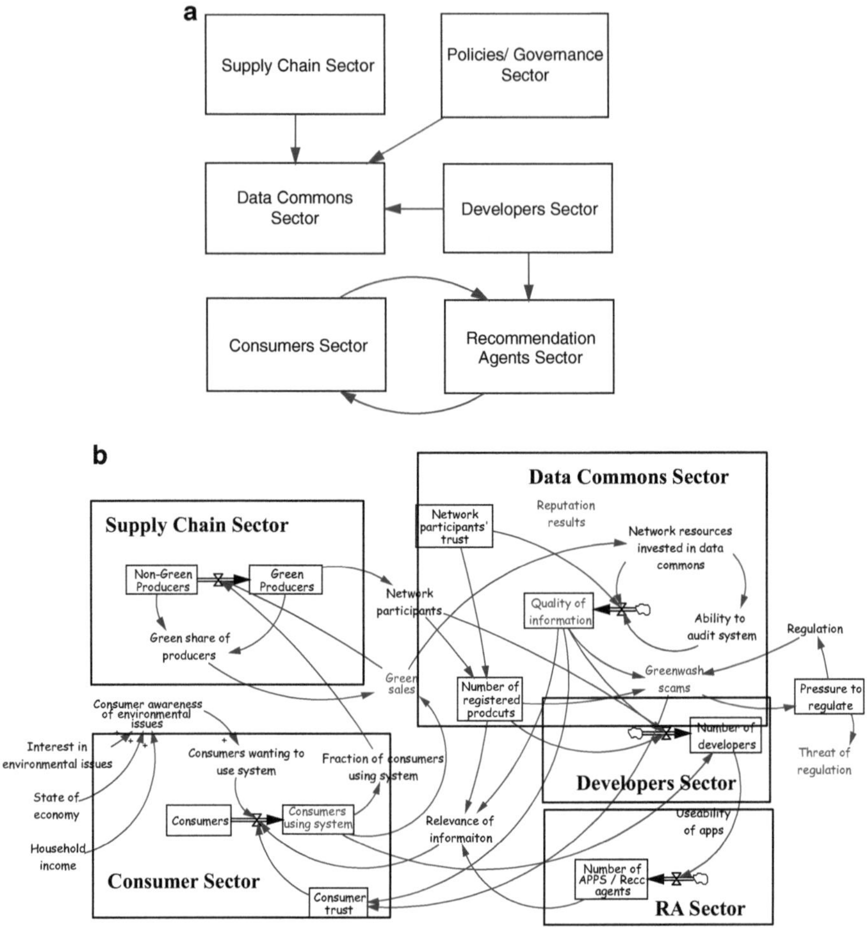

Fig. A.7 (**a**) A preliminary conceptual model from the GMB session. (**b**) Model sectors

Tinkering with formulations for where the needed resources to build capability would come from led us "back into" an assumed business model stating that producers (who would benefit financially from the operation of such a system) would supply the resources to construct system capability. Producer adoption was conceptualized to be the result of tracking the benefits and costs of the system, but in order to get serious about how such adoption could yield a resource base, we needed to expand those equations in the model.

And so step-by-step, the detail complexity of the proposed simulation model grew. But still the reformulated simulation seemed to be unwilling to create a self-sustaining takeoff. We had modeled consumer adoption as a more or less standard word-of-mouth innovation adoption with an initial "boost" from marketing. We discovered that a large enough marketing budget could force a consumer "takeoff," but without some large source of external resources, the model did not easily kick into a self-sustaining market growth. We decided to pay attention to the general

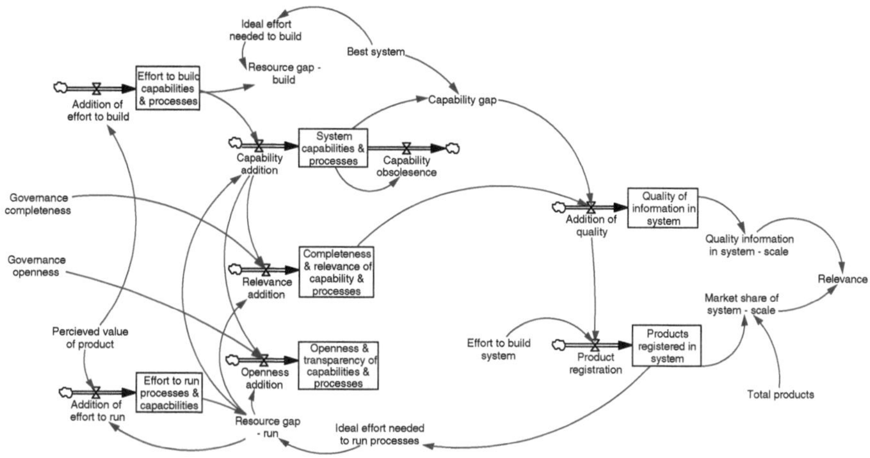

Fig. A.8 System map: sector I—Information Commons

admonition that "small models are beautiful models" (Ghaffarzadegan, Lyneis, & Richardson, 2010) and drop back and formulate a smaller reduced form model to get a better handle on overall model dynamics. The model is not reported in this book. However, the model is reported as a book chapter in a forthcoming title in the PAIT series (Ran et al., 2016).

A.4. I-Choose Prototype Design

A.4.1. Methods

We chose to develop our prototype as a semantic Web application. In order to develop ontologies and related semantic Web applications, the initial requirements are obtained through the development of use case scenarios (Gruninger & Fox, 1995). Such use cases were created to identify (1) the questions to be asked, (2) the resources that may be required to answer those questions, and (3) the methods by which to determine the answers (Fox & McGuinness, 2008). Table A.1, Step 1, outlines the process used to determine the primary question to be asked of our system: what constitutes a trustworthy certification and inspection process? To answer this question, we developed 28 questions that may be asked of certification and inspection data to qualify it as trustworthy. The questions were developed based on focus group discussions, interviews, and document analyses that the team conducted with stakeholders from the coffee supply chain including researchers, certifiers, members of producers' associations, retailers, and NGOs active in promoting sustainable and ethical trading as we described in the previous sections of this appendix.

Table A.1 Research methods and empirical evidence

Step	Description	Data collection method	Time and duration
1	Developing the use case	Focus group discussions with I-Choose network members	2011
		Interviews with producers, roasters, and certifiers	May to Aug 2012
		Document analysis	2012
		Develop 28 trust questions	
2	Map the structure of certification and inspection data	Interviews with certification body (Control Union, Fair Trade United Sates, and Fair for Life)	Nov 2012 to Feb 2013
		Archival analysis examining documentation of Flo-Cert	2012
		Mining data content from exemplar and audit report	2012
3	Developing ontology of certification	Focus group discussion with I-Choose network members	2011
		Interviews to refine the focus	May to Aug 2012
		Document analysis to identify semantic components	2012
4	Converting tabular data to triple data	Open source conversion tool csv2rdf4lod to convert the data format	2013
		OpenLink Virtuoso to convert into triple data	2013
5	Analysis of the 28 use case questions	SPARQL query	2013
		Inference-based retrieval of data	2013

The focus group was designed to gather information and gain an understanding of the full context of the sustainable certified coffee supply chain, starting from the coffee producers and ending with the coffee retailers. Following the focus group discussions, targeted interviews were conducted with the stakeholders of the sustainable certified coffee supply chain, including producers, roasters, and certifiers. The interview findings enabled the team to understand the challenges, information needs, and the process of certification and inspection for sustainable certified coffee. Finally, document analyses were conducted to complement the results found from the interviews. These methods enabled the team to identify 28 questions that characterize the trustworthiness of the certification and labeling schemes.

Based on the use case scenarios, we developed three ontologies. These three ontologies were further tested, verified, and validated following two approaches (Sayogo et al., 2016):

(a) Verification of ontology consistency. We used reasoners embedded in Protégé (Hermit ++ and Pellet) to verify our ontologies. To further evaluate the consistency of the ontologies, we created individual instances in them and reran the reasoner (Wang, Horridge, Rector, Drummond, & Seidenberg, 2005). The processes resulted in no inconsistencies in the ontologies.

(b) Validating competency questions using the proposed ontology. We also validated our ontologies by querying the data using competency questions. We queried using the DL query[1] plug-in available in Protégé and got satisfactory results.

A.4.2. How 28 Use Case Questions Analyze Data Trustworthiness

The last step in our method uses the 28 use case questions (Table A.2) to evaluate the overall trustworthiness of certification and inspection data. We use these 28 questions as a normative definition of trustworthiness. The certification trustworthiness evaluation system is based on two values, rightfulness and transparency of certification scheme. As such, our certification trustworthiness evaluation system consists of two major components: (a) trustworthiness evaluation criteria and (b) the data openness indicators. Ideal trustworthiness was assigned when all evaluation criteria were met. Inability to fulfill the criteria decreases the trustworthiness level.

The assessment of trustworthiness is based on matching the evaluation criteria with the degree of data openness, meaning the ease of extracting data to fulfill the evaluation criteria. The degree of data openness consists of two competing factors, data availability (data source) and governance level needed to extract the data. The less transparent the data is, the higher the governance level is needed to extract the data and vice versa. If the criteria cannot be met using these two indicators, it will decrease the trustworthiness level of the certification scheme. The measurement for these two indicators is listed in Tables A.3 and A.4. Data source (availability) measurement consists of five possible sources of data, as shown in Table A.3. There are three different levels of governance with specification listed in Table A.4. A more complete discussion of how each of the 28 questions was classified according to the three categories below can be found at https://github.com/jluciano/ichoose.

A.4.3. The Process of Examining CIDIBB

Generate a Fair-Trade Sample Data Set To test how CIDIBB could work, we mapped the structure of fair-trade certification and inspection data.[2] We created data tables to represent: (a) a certification body database structure that supports certification and inspection and (b) data aggregated by an information aggregator. We classified the data tables representing certification data as the certification and inspection database (CID) to refer to a database that consists only of certification and inspection results.

[1] DL query is query language that is based on the Manchester OWL syntax.

[2] These data are available in our open access repository (https://github.com/jluciano/ichoose). We published these datasets on http://ichoose.tw.rpi.edu/ with access through a SPARQL endpoint at http://ichoose.tw.rpi.edu/sparql

Table A.2 Twenty-eight use case questions

No.	Evaluation criteria	Assessment indicator	
		Data source (A ←→ F)	Governance (1 ←→ 3)
1	Is certification standard openly published (available on a website)		
2	Is certification compliance criteria/control points openly published (available on a website)		
3	Can know date of the inspection		
4	Can know date of certification		
5	Can know who is the inspector/auditor		
6	Can know how nonconformities are handled by the applicant		
7	Can know who is the standard-setting body		
8	Can know what type of organization made the standard (government, private, nonprofit)		
9	Can know who gives accreditation to the certifier		
10	Can know when the standard-setting body was established		
11	Is inspection report signed by an inspector		
12	Is certificate signed by the certifier		
13	Can know location of audit/inspection		
14	Is the list of nonconformity (measured score below standard) information available		
15	Can know the accreditation body of the standard-setting organization		
16	Can know the certification bodies of a particular standard-setting organization		
17	Can know the certification body of a particular applicant for particular products		
18	Does inspector/auditor have license		
19	Inspection/audit results openly published (available on request by FOIA or NGO)		
20	Does certification standard conform to a government-backed standard, e.g., USDA, EU-ECO-regulation		
21	Does certification standard conform to standard within an intergovernmental organization (e.g., ILO)		
22	Who sponsor the development of the standard (consumer NGO, producer, manufacture)		
23	Can know who translated the standard into compliance criteria/control points		
24	Is standard-setting body independent from the accreditation body, such as ISO		
25	Is certifier independent from the accreditation body		
26	Is certifier independent from the standard-setting body		
27	Is inspector/auditor independent from the standard-setting body		
28	Is inspector/auditor independent from certifier		

Table A.3 Data source (availability) levels

A	If data is available by searching the Web
B	If data is available in the certification and inspection database. This is the database available in the certifier system specifically to store information and data related to certification and inspection results
C	If data is available in regulator's database. The regulator here could be government agencies such as USDA (United State Department of Agriculture) or self-regulated organization such as ISO (International Standard Organization)
D	If data is available in the information system of the certifier but not in the certification and inspection database. An example of this database is the human resource database
E	If data is available in the database of a standard-setting organization. In a majority of certification schemes, the certifier is independent of a standard-setting organization
F	If the source of this data is not explicit and cannot be easily identified

Table A.4 Governance levels

1	There is no need to appeal to higher governance authority to access the data. This assumes that the data is available in the certification and inspection database and the certifier agrees to release the data
2	May need to appeal to higher governance authority to access the data because this data might exist in multiple data sources and one of the sources is outside of the certifiers' and standard-setting organization's information system
3	Higher governance intervention is needed to be able to answer the question because the source of the data is not explicit or cannot be easily identified

We created a total of eight data tables to represent the data structure of Flo-Cert, the main certification body under FLO. The data tables are contact data, product code, certification status, audit status, audit results, audit workflow status, corrective measure and objective evidence, and inspection checklist data. These eight data tables consist of 81 data attributes with each table comprising 8–13 attributes. Finally, two data tables with 14 data attributes were created to represent the list of data collected by an information aggregator. We also created two synthetic certification bodies that we named "Dave and Nic" certification body and "Nonviolent Dove" certification body. These two synthetic certification bodies were designed to follow practices less stringent than FLO practices and broadly congruent with "light greenwashing" (David and Nic) and "heavy greenwashing" (Nonviolent Dove).

In addition to the database described above, we manually created data tables representing the data structure that might be created by an information aggregator. Information aggregators in our ecosystem search the Web to extract data, then they refine and reformat the data for easy use. This dataset represents data about certification and standard bodies that are not within the more narrowly defined CID containing only information about direct certification activities. Hence information outside the CID contains data such as data tables which contain information about the governance and history of different certification and standard bodies.

Publish Data Set as a RDF Triple Store The data in our CID was formatted as standard tables of data such as those that might be found in a spreadsheet (such as EXCEL) or a relational database (such as ACCESS). Using the classes and relationships defined by the CerTIN and FLO ontologies, we used standard semantic Web technologies to recast those tables of data as an RDF triple store that is searchable using SPARQL queries.

Run SPARQL Queries Against the Data Testing the proof of concept begins by running a SPARQL query against the data in the triple store to see if the basic questions could be answered by such a direct query (see Fig. A.9). If the answer to the question could be retrieved using a SPARQL query or could be answered using the reasoning and inference tool, then we would examine the data source of the answer. Questions whose answers can be found directly in publicly available data or inferred by data provided by the certification and inspection agencies without any need to appeal to a higher authority were classified as "Level 1: No need to appeal for higher authority." For example, "Can I know the date of inspection?"

A second class of questions can still be answered either by a SPARQL query or by using the reasoning and inference tool, but the answer to these questions does not originate with the certifying and inspecting processes and organizations per se. These questions seek information about the certification and inspection organizations and processes themselves. Hence, answers to these questions require that data be made available in the triple store that refers to *some* higher authority. These questions were classified as "Level 2: May need appeal to a higher authority." For example, the question "Can I know the accrediting agency for the standard-setting body?" appeals to a higher authority to provide the name of the accrediting agency as this data is not available in the CID.

Finally, if questions could not be *answered* by a direct SPARQL query nor could be inferred using formal inference tools or manual curations of the data, these questions were classified as "Level 3: Requires governance intervention." For questions in this category, the data sources of their answers are not explicit, and governance interventions are required to locate the answer and make it publicly available. For example, the question "Is the inspector/auditor independent from the standard-setting body?" cannot be answered without intervention by some form of governance. We finally summarized how many of the 28 use case questions could be answered at each level.

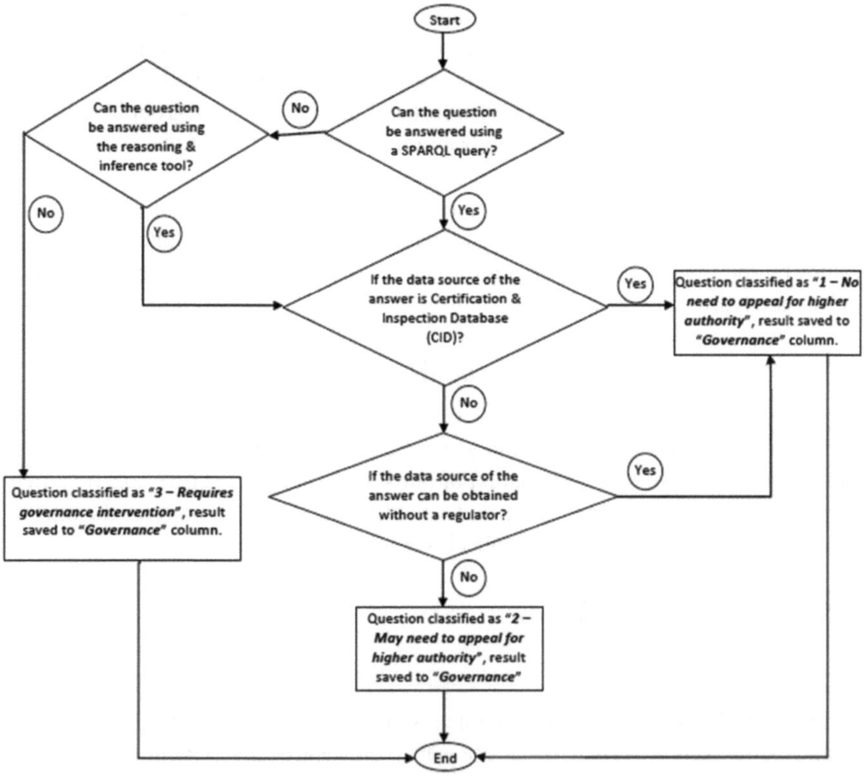

Fig. A.9 The process of examining CIDIBB usefulness in trustworthiness evaluation using 28 questions

A.5. Survey

In this section of the appendix, we introduce the methods and procedures that we used in the survey that we conducted to assess trustworthiness in information packages attached to labels. Again, although the results are not reported in this book, they informed our conclusions and are also in the process of being published elsewhere (Zhang, Liu, Sayogo, Picazo-Vela, & Luna-Reyes, Forthcoming).

A.5.1. Data Collection

We sought to assess the *empirical* relationships between six factors as independent variables and four control variables and the dependent variables in the form of trusting belief on sustainability claim. A survey was distributed to students enrolled in a private university in Mexico and a private university in the United States in September 2013 and a total of 178 responses were received. To provide some context for the research, a decision-making assignment was distributed among all students before answering the survey. The survey instrument was developed initially in English and revised by a panel of experts. It was applied in English and Spanish. The questionnaire was first translated to Spanish by a group of research assistants, and then it was translated back into English by two of the authors of the study to verify the accuracy of the translation.

After data cleaning, we excluded questionnaires with over 10 % missing values. Ten participants omit answering very few questions, and we substituted the missing values by the mean value of the missing item. At last, 167 observations were used for our analyses.

A.5.2. Variable Development and Measurement

All measurements used in this study were adapted from previous literature and adjusted to fit the purpose and context of our study. Table A.5 shows construct definitions, number of items measured, and reliability measures. In terms of the internal reliability estimation, we used Cronbach's α for all constructs.

A.5.3. Dependent Variables

Trusting Beliefs It consists of three sub-variables: competence, integrity, and benevolence. *Competence* measures the degree by which consumers believe that sustainable product labels include reliable and valid information that is appropriate for making purchase decisions. *Benevolence* measures the degree by which consumers believe that sustainable claim depicted in product label reflects the disposition of such claim to do something good and serve for the interests of consumers. *Integrity* measures the degree by which consumers believe that sustainable claim depicted in product package and label reflects truthfulness, honesty, and other integrity values.

A.5.4. Independent Variables

There are six independent variables of interest and four control variables with description as follows:

- *Disposition to trust label.* This variable measures the tendency of respondents to trust information provided by sustainable claim regardless of other reasons.
- *Brand and company reputation.* This variable measures the importance of information about product brand and company's reputation for consumers.
- *Certification reputation.* This variable measures the importance of reputation of the sustainable certification scheme for consumers.

Table A.5 Construct definitions and items measured

Constructs	Definitions of this study	No. of items	References	α
Dependent variable				
Competence (COM)	Information depicted in product labels is capable of guiding consumers to make purchasing decisions	4	Porter and Donthu (2008)	0.76
Benevolence (BENE)	Information depicted in product labels reflects the consideration of welfare and interests of consumers	3	Porter and Donthu (2008)	0.72
Integrity (INT)	Information depicted in product label reflects truthfulness, honesty, and upholding of other ethical values	3	Porter and Donthu (2008)	0.79
Independent variable				
Brand and company reputation (BACR)	The importance of product brand and company's reputation	2	Grabner-Kraeuter & Kaluscha (2003), De Pelsmacker, Driesen, & Rayp (2005)	0.80
Certification reputation (CR)	The importance of reputation of the sustainable certification scheme	3	Jiang et al. (2008), Jøsang et al. (2007), Grabner-Kraeuter & Kaluscha (2003)	0.54
Additional information to verify label (AI)	The importance of additional information to verify label	2	McKnight, Choudhury, and Kacmar (2002), Vance, Elie-Dit-Cosaque, and Straub (2008)	0.60
Government endorsement (GE)	The importance of government endorsement in terms of legal and regulatory system and information source	2	Li, Hess, and Valacich (2008), McKnight et al. (2002)	0.68
NGO-based label (NGO)	The importance of label supported or developed by NGOs	2	Michaelidou & Hassan (2010)	0.71
Disposition to trust label (DTT)	The tendency of respondents to trust information provided by sustainable certification scheme regardless of other reasons	3	Gefen (2000), Jiang, Jones and Javie (2008)	0.79

(continued)

Table A.5 (continued)

Constructs	Definitions of this study	No. of items	References	α
Control variable				
Prior knowledge (PR)	The extent of respondents' prior knowledge of sustainable certification and labeling practices in general	3	Gleim, Smith, Andrews, et al. (2013)	0.83
Country (COU)	The country of birth of the respondent	1	N/A	N/A
Gender (GEN)	The gender of the respondent	1	N/A	N/A
Education level (EDU)	The education level of the respondent	1	N/A	N/A

- *Additional information to verify label.* This variable measures to what extent the existence of additional information to verify label is important for consumers.
- *Government endorsement.* This variable measures the importance of endorsement by government in terms of legal and regulatory support and information source for consumers.
- *NGO-based label.* This variable measures the importance of label supported or developed by NGOs for consumers.
- *Prior knowledge of label.* This is a control variable measuring the extent of respondents prior knowledge of sustainable certification and labeling practices in general.
- *Country.* This is a control variable indicating the country of birth of the respondent. We divided the country into three—US-born respondents, Mexican-born respondents, and other countries-born respondents.

In addition, in order to ensure the differences were not caused by gender and education level (graduate vs. undergraduate) of the sample, we use gender and level of education also as control variables.

A.5.5. Survey Instrument

Dear student,

We are currently conducting a research project funded by a grant from the National Science Foundation titled "Building Information Sharing Networks to Support Consumer Choice," or "I-Choose" for short. The main goal of the I-Choose project is to develop interoperable information system to support sustainable consumption, using a case of coffee produced in Mexico and sold in the United States and Canada.

We have found that a key component of the project is related to consumer trust. We prepared this survey to better understand your position about a series of statements related to trust in product information, including elements present on product labels.

The survey will take approximately 15 min. Your participation is completely voluntary and you are free to quit the survey at any time or to not answer individual questions. Should you decide to participate, all information will remain anonymous. The survey asks only for some demographic information, and we will present only aggregated data in any publications based on the data. Although there will be no direct benefits to you, this research will help to inform others about the role that trust plays in creating product packaging. There are no risks in answering the survey, and your answers will not affect your grades.

If you have any questions or are interested in knowing the results, please get in touch with any of us. Thank you for participating in our research project.

Sincerely,

Jing Zhang, Sergio Picazo-Vela, and Luis F. Luna-Reyes

Background Information

In today's market, companies use product certification and labels, as well as information in product packaging to convey product quality and inspire trust from consumers. Product certification refers to the label or logo endorsing that production, and distribution practices might comply with a specific standard that avoids some negative externalities, such as child labor, non-fair trading, and non-environmentally friendly practices. Product certification is organized into three approaches, 1st party, 2nd party, and 3rd party certification. The difference between these certifications is the degree of separation between the certifier and the company whose product is being certified.

- First-party certification means self-declared adherence to voluntary policies or codes of conduct that emerge inside the company. An example of first-party certification is Nespresso AAA (Fig. A.11a).

Fig. A.11 Kinds of product certifications. (**a**) First-party certification: Nespresso AAA, a self-declared adherence to Green practices by Nestle. (**b**) Second-party certification: SCAA sets the standards and certifies coffee brewing equipment. (**c**) Third-party certification: FLO labeling international, nonprofit-based third-party certification

Fig. A.12 Product information in the product packaging

- In second-party certification, standards and certificates are issued by an industry or trade association and adopted by firms. An example of second-party certification is SCAA certifying coffee brewing equipment (Fig. A.11b).
- Third-party certification is when one organization sets the standard, and compliance to the standard is tested and awarded by an independent third-party certifier, which audits firms that want to adopt the standard. An example of third-party certification is FLO labeling international (Fig. A.11c).

In addition to the adoption of one or more certificates, companies also provide additional information in their product packaging to tell stories, including their production practices or additional efforts to promote sustainable practices and trading. As an example, Fig. A.12 demonstrates how equal exchange uses its product packaging to convey a variety of information.

Instructions

Please read and answer the following questions. There are no correct or incorrect answers; we only wish to learn more about your beliefs and previous experience related to product information.

A. *General questions/previous experience*

1. How familiar are you with the following certificate labels?

Label	Never have seen it			I know it very well	
FAIRTRADE	1	2	3	4	5
CERTIFIED	1	2	3	4	5
FAIR TRADE CERTIFIED	1	2	3	4	5
UTZ CERTIFIED Good inside	1	2	3	4	5
BIRD FRIENDLY	1	2	3	4	5
USDA ORGANIC	1	2	3	4	5
4C ASSOCIATION	1	2	3	4	5

2. I buy sustainable/certified products.
   ☐ Never ☐ Rarely ☐ Sometimes ☐ Frequently ☐ Always
3. I use an information aggregator website to make buying decisions (e.g. GoodGuide, Barcoo).
   ☐ Never ☐ Rarely ☐ Sometimes ☐ Frequently ☐ Always
4. I am very self-conscious about my health when buying products and services.
   ☐ Strongly Agree ☐ Agree ☐ Neither agree nor disagree ☐ Disagree ☐ Strongly Disagree
5. I am very self-conscious about sustainability when buying products and services.
   ☐ Strongly Agree ☐ Agree ☐ Neither agree nor disagree ☐ Disagree ☐ Strongly Disagree
6. I consider myself an expert on identifying sustainable product labels and certifications.
   ☐ Strongly Agree ☐ Agree ☐ Neither agree nor disagree ☐ Disagree ☐ Strongly Disagree
7. I have a great deal of knowledge about sustainable product labels and certifications.
   ☐ Strongly Agree ☐ Agree ☐ Neither agree nor disagree ☐ Disagree ☐ Strongly Disagree
8. I generally know more than my friends about sustainable product labels and certification.
   ☐ Strongly Agree ☐ Agree ☐ Neither agree nor disagree ☐ Disagree ☐ Strongly Disagree

B. *General belief.* Please indicate to what degree you agree with the following statements.

1. I generally trust other people.

☐ Strongly Agree ☐ Agree ☐ Neither agree nor disagree ☐ Disagree ☐ Strongly Disagree

2. I generally trust information from government agencies.

☐ Strongly Agree ☐ Agree ☐ Neither agree nor disagree ☐ Disagree ☐ Strongly Disagree

3. I generally trust information from private companies.

☐ Strongly Agree ☐ Agree ☐ Neither agree nor disagree ☐ Disagree ☐ Strongly Disagree

4. I generally trust information provided by non-government organizations.

☐ Strongly Agree ☐ Agree ☐ Neither agree nor disagree ☐ Disagree ☐ Strongly Disagree

5. I generally have faith in third party certifications.

☐ Strongly Agree ☐ Agree ☐ Neither agree nor disagree ☐ Disagree ☐ Strongly Disagree

6. I generally trust information supplied by product labels and packaging

☐ Strongly Agree ☐ Agree ☐ Neither agree nor disagree ☐ Disagree ☐ Strongly Disagree

7. I tend to count upon information supplied by product labels and packaging

☐ Strongly Agree ☐ Agree ☐ Neither agree nor disagree ☐ Disagree ☐ Strongly Disagree

8. I generally have faith in humanity

☐ Strongly Agree ☐ Agree ☐ Neither agree nor disagree ☐ Disagree ☐ Strongly Disagree

9. I generally trust information supplied by product labels and packaging unless they give me reason not to.

☐ Strongly Agree ☐ Agree ☐ Neither agree nor disagree ☐ Disagree ☐ Strongly Disagree

C. *Antecedents of trust.* The following statements measure sets of conditions that could influence how you trust information detailing sustainable practice (adherence to economic, social, and environmental values). Please indicate to what degree you agree with the following statements.

1. The existence of any logo or label of third-party certification promoting sustainable practice is very important for my decision.
☐ Strongly Agree ☐ Agree ☐ Neither agree nor disagree ☐ Disagree ☐ Strongly Disagree

2. I tend to ignore information about sustainable and ethical practices in the product packaging if I see any label or logo from third-party certification.
☐ Strongly Agree ☐ Agree ☐ Neither agree nor disagree ☐ Disagree ☐ Strongly Disagree

3. It is important to me to have additional information about the certification process such as dates of inspection or the name of the inspector.
☐ Strongly Agree ☐ Agree ☐ Neither agree nor disagree ☐ Disagree ☐ Strongly Disagree

4. It is important to me to be able to verify certification information over the Internet or other apps.
☐ Strongly Agree ☐ Agree ☐ Neither agree nor disagree ☐ Disagree ☐ Strongly Disagree

5. Seeing a label from a well-known third-party certifier promoting sustainable practice assures me that the product adheres to promoting sustainable and ethical trading.
☐ Strongly Agree ☐ Agree ☐ Neither agree nor disagree ☐ Disagree ☐ Strongly Disagree

6. A label from a well-known third-party certifier contributes to a lessening of my perceived risk of not getting what I am paying for.
☐ Strongly Agree ☐ Agree ☐ Neither agree nor disagree ☐ Disagree ☐ Strongly Disagree

7. Seeing a label from a lesser-known third-party certifier increases my doubt and risks associated with the adherence to promote sustainable and ethical trading.
☐ Strongly Agree ☐ Agree ☐ Neither agree nor disagree ☐ Disagree ☐ Strongly Disagree

8. A self-declared label from a company (first party label) increases my suspicion and perceived risks associated with promotion of sustainable and ethical trading.
☐ Strongly Agree ☐ Agree ☐ Neither agree nor disagree ☐ Disagree ☐ Strongly Disagree

9. If the product is produced by a well-known seller/manufacturer, third-party certification becomes less important for my decision
☐ Strongly Agree ☐ Agree ☐ Neither agree nor disagree ☐ Disagree ☐ Strongly Disagree

10. If the brand of the product is well-known, third-party certification becomes less important to my decision
☐ Strongly Agree ☐ Agree ☐ Neither agree nor disagree ☐ Disagree ☐ Strongly Disagree

11. Pictures from producers or maps showing product origin included in the product packaging help my decision to buy a product.
☐ Strongly Agree ☐ Agree ☐ Neither agree nor disagree ☐ Disagree ☐ Strongly Disagree

12. Text describing adherence to sustainable and ethical practices on the product packaging is useful in making buying decisions.
☐ Strongly Agree ☐ Agree ☐ Neither agree nor disagree ☐ Disagree ☐ Strongly Disagree

13. I pay attention to specific product contents or ingredients provided in the product packaging before making buying decisions.
☐ Strongly Agree ☐ Agree ☐ Neither agree nor disagree ☐ Disagree ☐ Strongly Disagree

14. A link on the product package to a company's website describing its sustainable and ethical practices is useful in making buying decisions.
☐ Strongly Agree ☐ Agree ☐ Neither agree nor disagree ☐ Disagree ☐ Strongly Disagree

15. I feel secure in relying on product labels and certifications by the government because legal systems exist to protect me from any falsification or forgery.
☐ Strongly Agree ☐ Agree ☐ Neither agree nor disagree ☐ Disagree ☐ Strongly Disagree

16. A link on a product label to a government website verifying sustainable and ethical practices and certification is useful in making buying decisions.
☐ Strongly Agree ☐ Agree ☐ Neither agree nor disagree ☐ Disagree ☐ Strongly Disagree

17. I feel secure in relying on product labels and certifications by NGOs such as Fairtrade or Rainforest Alliance.
☐ Strongly Agree ☐ Agree ☐ Neither agree nor disagree ☐ Disagree ☐ Strongly Disagree

18. A link on a product label to an NGO's website verifying sustainable and ethical practices and certification is useful in making buying decisions.
☐ Strongly Agree ☐ Agree ☐ Neither agree nor disagree ☐ Disagree ☐ Strongly Disagree

19. Product labels and certifications recommended by my friends are important for my buying decisions.
☐ Strongly Agree ☐ Agree ☐ Neither agree nor disagree ☐ Disagree ☐ Strongly Disagree

20. Product labels and certifications recommended by my family are important for my buying decisions.
☐ Strongly Agree ☐ Agree ☐ Neither agree nor disagree ☐ Disagree ☐ Strongly Disagree

D. *Trusting belief* The following statements measure your beliefs about trusting information contained in a product's label and package that refers to sustainable practices in producing the product. Please indicate to what degree you agree with the following statements.

1. I trust that sustainable product labels and packaging include information that is appropriate for me to make buying decisions.
☐ Strongly Agree ☐ Agree ☐ Neither agree nor disagree ☐ Disagree ☐ Strongly Disagree

2. I trust that sustainable product labels and packaging seldom include information that is misguiding.
☐ Strongly Agree ☐ Agree ☐ Neither agree nor disagree ☐ Disagree ☐ Strongly Disagree

3. I trust that sustainable product labels and packaging include reliable information about sustainable practices.
☐ Strongly Agree ☐ Agree ☐ Neither agree nor disagree ☐ Disagree ☐ Strongly Disagree

4. I trust that sustainable product labels and packaging include valid information about sustainable practices.
☐ Strongly Agree ☐ Agree ☐ Neither agree nor disagree ☐ Disagree ☐ Strongly Disagree

5. I trust that sustainable product labels and packaging include information that considers my welfare.
☐ Strongly Agree ☐ Agree ☐ Neither agree nor disagree ☐ Disagree ☐ Strongly Disagree

6. I trust that sustainable product labels and packaging include information with my best interests in mind.
☐ Strongly Agree ☐ Agree ☐ Neither agree nor disagree ☐ Disagree ☐ Strongly Disagree

7. I trust that sustainable product labels and packaging include information related to my interests and values.
☐ Strongly Agree ☐ Agree ☐ Neither agree nor disagree ☐ Disagree ☐ Strongly Disagree

8. I trust that sustainable product labels and packaging include honest and truthful information.
☐ Strongly Agree ☐ Agree ☐ Neither agree nor disagree ☐ Disagree ☐ Strongly Disagree

9. I trust that sustainable product certifications ensure a set of standards about sustainable practices.
☐ Strongly Agree ☐ Agree ☐ Neither agree nor disagree ☐ Disagree ☐ Strongly Disagree

10. I trust that sustainable product certifications can be counted on representing the right sustainable practices.
☐ Strongly Agree ☐ Agree ☐ Neither agree nor disagree ☐ Disagree ☐ Strongly Disagree

E. *Demographic questions*

1. Please indicate your gender.
☐ Female ☐ Male ☐ Prefer not to answer

2. Please indicate your current educational status.
☐ I am an undergraduate student.
☐ I am a graduate student.

3. Please indicate the country where you were raised ____________________

4. Are you currently employed part-time, full-time, or not employed?
☐ Part-time ☐ Full-time ☐ Not employed ☐ Prefer not to answer

5. Please indicate your age.
☐ 17 – 20 years ☐ 21 – 25 years ☐ 26 – 30 years
☐ 31 – 35 years ☐ 36 – 40 years ☐ > 40 years

References

Black, L. J., & Andersen, D. F. (2012). Using visual representations as boundary objects to resolve conflict in collaborative model-building approaches: Using visual representations as boundary objects. *Systems Research and Behavioral Science, 29*(2), 194–208. http://doi.org/10.1002/sres.2106.

De Pelsmacker, P., Driesen, L., & Rayp, G. (2005). Do consumers care about ethics? Willingness to pay for fair-trade coffee. *Journal of Consumer Affairs, 39*(2), 363–385.

Fox, P., & McGuinness, D. L. (2008). *TWC semantic web methodology*. Retrieved from http://tw.rpi.edu/web/doc/TWC_SemanticWebMethodology.

Gefen, D. (2000). E-commerce: The role of familiarity and trust. *Omega, 28*(6), 725–737. doi:10.1016/S0305-0483(00)00021-9.

Ghaffarzadegan, N., Lyneis, J., & Richardson, G. P. (2010). How small system dynamics models can help the public policy process. *System Dynamics Review*, n/a–n/a. http://doi.org/10.1002/sdr.442.

Gleim, M. R., Smith, J. S., Andrews, D., & Cronin, J. J., Jr. (2013). Against the green: A multi-method examination of the barriers to green consumption. *Journal of Retailing, 89*(1), 44–61. doi:10.1016/j.jretai.2012.10.001.

Grabner-Kräuter, S., & Kaluscha, E. A. (2003). Empirical research in on-line trust: A review and critical assessment. *International Journal of Human-Computer Studies, 58*(6), 783–812.

Gruninger, M., & Fox, M. S. (1995). Methodology for the design and evaluation of ontologies. In *Proceedings of the workshop on basic ontological issues in knowledge sharing*. Montreal.

Jiang, P., Jones, D. B., & Javie, S. (2008). How third-party certification programs relate to consumer trust in online transactions: An exploratory study. *Psychology and Marketing, 25*(9), 839–858.

Jøsang, A., Ismail, R., & Boyd, C. (2007). A survey of trust and reputation systems for online service provision. *Decision Support Systems, 43*(2), 618–644.

Komiak, Y. X. K., & Benbasat, I. (2006). The effects of personalization and familiarity on trust and adoption of recommendation agents. *MIS Quarterly, 30*(4), 941–960.

Li, X., Hess, T. J., & Valacich, J. S. (2008). Why do we trust new technology? A study of initial trust formation with organizational information systems. *The Journal of Strategic Information Systems, 17*(1), 39–71. doi:10.1016/j.jsis.2008.01.001.

Luna-Reyes, L. F., Zhang, J., Roy, R., Andersen, D. F., Whitmore, A., & Andersen, D. L. (2013). Information strategies to support full information product pricing: The role of trust. *Information Polity, 18*(1), 75–91.

McKnight, D. H., Choudhury, V., & Kacmar, C. (2002). Developing and validating trust measures for e-Commerce: An integrative typology. *Information Systems Research, 13*(3), 334–359.

Michaelidou, N., & Hassan, L. M. (2010). Modeling the factors affecting rural consumers purchase of organic and free-range produce: A case study of consumers from the Island of Arran in Scotland, UK. *Food Policy, 35*(2), 130–139.

Porter, C. E., & Donthu, N. (2008). Cultivating trust and harvesting value in virtual communities. *Management Science, 54*(1), 113–128. doi:10.1287/mnsc.1070.0765.

Ran, W., Jarman, H., Luna-Reyes, L. F., Zhang, J., Andersen, D. L., Tayi, G. K., et al. (2016). Supply-chain transparency and governance systems: Market penetration of the I-choose system. In J. Zhang, L. F. Luna-Reyes, T. A. Pardo, & D. S. Sayogo (Eds.), *Information, models, and sustainability: Policy informatics in the age of big data and open government*. Springer. Basel, Switzerland.

Sayogo, D. S., Ran, W., Tayi, G. K., Luciano, J. S., Luna-Reyes, L. F., Depaula, D., et al. (2016). Ontological modeling of certification and inspection process to support smart disclosure of product information. *International Journal of Public Administration in the Digital Age 3*(2), 87–109.

Sayogo, D. S., Zhang, J., Luna-Reyes, L., Jarman, H., Tayi, G., Andersen, D. L., et al. (2015). Challenges and requirements for developing data architecture supporting integration of sustainable supply chains. *Information Technology and Management, 16*(1), 5–18. http://doi.org/10.1007/s10799-014-0203-3.

Vance, A., Elie-Dit-Cosaque, C., & Straub, D. W. (2008). Examining trust in information technology artifacts: The effects of system quality and culture. *Journal of Management Information Systems, 24*(4), 73–100.

Wang, H., Horridge, M., Rector, A., Drummond, N., & Seidenberg, J. (2005). Debugging OWL-DL ontologies: A heuristic approach. In Y. Gil, E. Motta, V. R. Benjamins, & M. A. Musen (Eds.), *The semantic web – ISWC 2005* (pp. 745–757). Berlin/Heidelberg: Springer. Retrieved from http://link.springer.com/chapter/10.1007/11574620_53.

Yin, R. (1994). *Case study research: Design and methods*. Newbury Park: Sage.

Zhang, J., Liu, H., Sayogo, D. S., Picazo-Vela, S., & Luna-Reyes, L. F. (Forthcoming). Strengthening institutional-based trust for sustainable consumption: Lessons for smart disclosure. *Government Information Quarterly*.

Index

H. Jarman, L.F. Luna-Reyes (eds.), *Private Data and Public Value*,
Public Administration and Information Technology 26,
DOI 10.1007/978-3-319-27823-0

MIX
Papier aus verantwortungsvollen Quellen
Paper from responsible sources
FSC® C105338

If you have any concerns about our products,
you can contact us on
ProductSafety@springernature.com

In case Publisher is established outside the EU,
the EU authorized representative is:
Springer Nature Customer Service Center GmbH
Europaplatz 3, 69115 Heidelberg, Germany

Printed by Libri Plureos GmbH
in Hamburg, Germany